Spinning Our Wheels

About The Arthur Morgan Institute for Community Solutions

The Arthur Morgan Institute for Community Solutions (a.k.a. Community Solutions) is a non-profit organization based in Yellow Springs, Ohio that provides knowledge and practices on low-energy living. Our primary focus is on reducing energy consumption in the household sectors of housing, transportation and food (67% of total energy consumed in the U.S.). Community Solutions provides clear, high-quality data that helps people gain the analytical skills required to evaluate both the personal and societal options for reducing energy use. Founded in 1940, Community Solutions has had a long history of promoting small, local community values in both towns and neighborhoods.

Since 2003, Community Solutions has organized five national conferences on solutions for Peak Oil and climate change, written 19 New Solutions white papers, wrote the book *Plan C: Community Survival Strategies for Peak Oil and Climate Change* (New Society, 2008), produced the book *The Green Tragedy: LEEDs Lost Decade*, given more than 150 presentations at conferences and events and produced the award-winning film *The Power of Community: How Cuba Survived Peak Oil* (2006). Since 2005 we have worked closely with other transportation people in our local community as well as internationally.

Spinning Our Wheels

The Failures of the Auto Industry and Government in the Quest for Lower Carbon Emissions, and What We Can Do – Right Now – to Change Our Transportation System

by Pat Murphy

The Arthur Morgan Institute
for Community Solutions
Yellow Springs, Ohio

Printed in U.S.A.
ISBN:

To order or for inquiries regarding requests to reprint all or part of *Spinning Our Wheels* contact Community Solutions at the address below.

Arthur Morgan Institute for
Community Solutions
PO Box 243
114 E. Whiteman Street
Yellow Springs, Ohio 45387
937-767-2161

info@communitysolution.org
www.communitysolution.org
www.powerofcommunity.org
www.greentragedy.org

Contents

Acknowledgements

I received significant help in development of this book by a few people who actually think a paradigm shift in the mode of personal travel is a good idea. In all the myriad ways being suggested to stop climate change and deal with rapidly dwindling oil resources, only a small number of people consider the possible abandonment of the personal car. Fortunately, those individuals crossed my path.

Jan Lundberg, author and activist, first suggested in his magazine *Auto Free Times* that the private car might not be the blessing that the vast majority of people believe it to be. His work over the years and his lifestyle shows that giving up the personal car can be done.

Bob Steinbach of the Miami Valley Regional Planning Commission in Dayton, Ohio approached me after reading my first paper on the Smart Jitney. He later spoke at one of Community Solutions' conferences and has participated in studies that I did on the idea of a very different transport mode. His critiques have been very helpful.

Frank Goetzke, economics professor at the University of Kentucky, also critiqued the idea and offered a vision that melded the Smart Jitney with other innovative forms of transport that could be used within our home town of Yellow Springs, Ohio.

Chandra Vikash, an innovative thinker from India, spent a week with me looking at the Smart Jitney as well as acquainting me with his well thought out innovative vision, which he labeled MetroLITE, for a future that incorporated the Jitney as well as other alternative transport modes.

Many thanks to Sean O'Sullivan, president of the company Avego in Kinsale, Ireland, who has implemented a real life version of the Smart Jitney. His work has verified the concept and shown that it can be implemented relatively quickly.

Faith Morgan, Megan Quinn Bachman and Rob Content all contributed significantly to this work with their ideas and editing skills. John Rockwell,

Jeanna Breza and Wanda Rockwell did an excellent job of proofreading the final book.

John Morgan did the layout and book design. Bob Bingenheimer created the graphics, cover and index.

I acknowledge the people of Cuba who, during their energy crisis which they call The Special Period, developed a wide variety of shared transport options which first generated this idea in my mind. I am also grateful to other Third World countries who are modeling this kind of system in a less sophisticated but viable form.

Preface

When I began studying Peak Oil in 2003 I expected the world would suffer several major inconveniences, but there was little doubt in my mind that the challenges could be met. Al Gore's 2006 film "An Inconvenient Truth" and the 2007 Fourth Assessment Report by the Intergovernmental Panel on Climate Change changed my mind – things were much more serious than I had thought and a rapid reduction in CO_2 generation became an imperative.

In 2007 I proposed the "Smart Jitney – Rapid Realistic Transport" in our *New Solutions* report number 12. It had become clear to me that continued reliance on the private car as the chief mode of transportation was probably not sustainable. In the next two years the Smart Jitney idea was proposed by many different people – it is not a particularly original idea. And it has been implemented in a variety of forms in Third World countries. Cell phones and Global Positioning Systems (GPS) combined with mapping software make it much easier to automate the process, adding a level of convenience that will make it more attractive to Americans.

In 2008 I learned about an Irish company Avego that was developing software for shared transport and began communicating with Avego's president, Sean O'Sullivan. The idea continued to ferment around the world and the first conference, entitled "Real-Time Rides: A Smart Roadmap to Energy and Infrastructure Efficiency" was put on by the MIT Real-time Rides Research group in April, 2009. Rob Content, program manager for Community Solutions, spoke at the meeting. Many useful papers and ideas came from the participants, one who traveled from New Zealand. This conference showed that the idea was being considered around the world. In May of 2009, Sean demonstrated Avego's shared transport system to me at his headquarters in Kinsale, Ireland.

During this period, the Pluggable Hybrid Electric Vehicle (PHEV) continued to be marketed extensively. Bold claims were made about its performance, none verifiable as this book will show. In August 2009, the president of GM, Fritz Henderson, announced that the Volt would achieve city fuel economy of at least 230 miles per gallon. Not to be outdone, Nissan quickly announced that its new Leaf EV would average 367 mpg. Shortly thereafter the EPA announced that it had no way of computing the miles per gallon for a Volt but that it supported the development of the car as a great American idea.

Statements like these are what led me to write this book. I am not against any automobile development that will reduce CO_2 emissions and the consumption of oil. But the claims made for these new products imply revolutionary breakthroughs that eliminate any need to actually reduce consumption as measured by number of cars or miles driven. It allows the continuation of the dream of a car (or two or three) in every garage in every household in the world. And this dream has played and continues to play a significant part in the endangerment of life on earth. Governments support the dream since in industrialized countries the private car is essential to continued economic growth.

In addition, the idea of an electric car, pluggable or not, is a shift away from burning petroleum as gasoline or diesel in a vehicle to burning coal, natural gas, and uranium in a power plant, which would send electricity via the power grid to hundreds of millions of batteries in hundred of millions of cars. Thus CO_2 and other pollutants would be generated, not from the hundreds of millions of cars, but from tens of thousands of power plants. Most of the power plants will burn coal, mined with devastating ecological impacts. This will not reduce CO_2 emissions 80-90% by 2050 as called for by the IPCC and now most major governments. Naturally the proponents of electric cars suggest burial of power plant CO_2 emissions as well as electricity from wind and solar photovoltaics (PV) to address this concern. But I am not convinced that burying CO_2 is technically feasible, given the enormous volumes and the rudimentary state of the technology. Although such claims may be achievable, the risk is great.

Shared transport, as described in Part 3 of this book, is a very inexpensive alternative. It has been implemented by one small company so far with a very small investment when compared to the billions of dollars of investment in electric cars of all kinds. It provides a contingency plan, a backup option. And it provides an option for Third World countries to improve their national transportation without committing to the private car.

For decades, U.S. car companies, in collaboration with the U.S. government, have proposed radical technologies to replace the gasoline car. In 1970 President Nixon said, "I am inaugurating a program to marshal both government and private research with the goal of producing an unconventionally powered virtually pollution free automobile within five years." When launching Project Independence three years later he noted, "Let this be our national goal: At the end of this decade, in the year 1980, the United States will not be dependent on any other country for the energy we need to provide our jobs, to heat our homes, and to keep our transportation moving." Since then car companies have tried one technology after the other, beginning with the fuel cell car, followed by the electric car, then diesel hybrids, back to the fuel cell idea again, and now focusing on the PHEV. This book explores that checkered unprofitable past and suggests that a clean super-efficient private car may never be realized. Shared transport, identified with the label "Smart Jitney," is an idea which needs to be seriously considered. The purpose of this book is to advocate for that consideration.

Part 1
The History of Electric Cars

Introduction

Americans fervently believe in technology, concluding that it is fundamentally good and one of the main reasons for America's greatness. Many believe technology improvements, or "techno-fixes," will save us from the twin threats of Peak Oil and climate change, both essentially the result of earlier technologies. When technology causes environmental damage and resource depletion, the American response is to strive for new, supposedly more benign, technology. Nowhere is this truer than in the case of the private automobile, possibly the most destructive technology ever invented (and the one most loved by the average citizen). As carbon dioxide (CO_2) levels head toward an atmospheric point of no return, Americans continue their so-called love affair with the automobile.

The focus on techno-fixes enables the avoidance of effective actions that could mitigate resource depletion and CO_2 generation. For example, the hope for "super-efficient cars" acts as a constant barrier to improvements in public transportation and other forms of shared transit. The 70 million new cars produced in the world each year will generate CO_2 for 14 years (the average automobile life span) and seriously damage the planet. Making the automobile somewhat more efficient will not stop the damage.

The current leading techno-fix product is the pluggable hybrid electric vehicle (PHEV), lauded by utilities, coal companies, car companies, activists, greens, environmentalists, liberals, and conservatives alike. President Obama has committed the United States to building one million pluggable hybrids, averaging 150 miles per gallon (mpg), by 2015.[1] The underlying hope is that the PHEV, once perfected, will somehow solve our energy supply problems and clean up the environment without any major inconvenience to the driving public.

The enthusiasm for the PHEV is based principally on highly misleading mpg claims. In August 2009 the new president of General Motors (GM), Fritz Henderson, provided such a claim for the Chevrolet Volt PHEV: 230 miles per gallon![2] Tragically, these numbers are bogus. To counter the propaganda of car, coal, and power companies, an "apples to apples" comparison is needed. It is important to bring the current misleading mpg comparison for the PHEV and other battery-only cars to scientific measures such as CO_2 generated or BTUs (British thermal units) consumed per mile traveled. This book brings a badly needed perspective to the excessive performance claims of all kinds of electric vehicles.

What's New Is Old

It is important to understand the history of pure battery electric vehicles (BEVs) as well as more recent proposals for hybrid electric vehicles (HEVs), which use both gasoline and batteries. Surprisingly, the energy technologies on which we are placing our hopes for new car breakthroughs are relatively old.[3] In 1900 more electric cars were sold than gasoline powered vehicles! The gasoline engine superseded the electric motor but the technology never disappeared and electric vehicles in some form or other (like golf carts and fork lifts) have been in continuous use for decades. Even the fuel cell electric car has a long history. (The hydrogen in the fuel tank of a fuel cell car generates electricity to operate an electric drive train.) For example, in 1903 in Denmark, Poul la Cour developed a system using a wind turbine to generate electricity to remove hydrogen from water. The hydrogen was then used to heat the schoolhouse. Thus began the "hydrogen economy," a term first used in 1970,[4] based on the concept of the fuel cell, a device first built in 1839. The electric car in all its modern manifestations is based on principles that are decades old.

The 1970s – The Birth of the Hybrid

The PHEV is not as new a concept as is popularly thought, having originated in the 1970s. In 1972, scientist Victor Wouk concluded that battery electric vehicles were not commercially viable due to battery limitations. He began thinking of combining the low-emissions benefits of an electric car with the power of a gasoline engine to produce an electric hybrid vehicle (one that runs mostly from the charged batteries). Wouk was given a government contract to build a prototype. He combined a 1972 Buick Skylark body, lead acid batteries, and a Mazda Wankel engine to build an electric hybrid vehicle, which he completed in 1974.[5] Although it passed EPA emissions tests, further funding did not materialize.

Wouk has been called the "father of the hybrid car," but possibly the title is not completely accurate. His car batteries were to be charged from the electrical grid, with the Wankel engine serving as a backup gasoline-powered generator. Surprisingly, this configuration is very similar to proposed new designs such as that of the Chevrolet Volt. Wouk's system was not an early gasoline hybrid car, as implemented by Toyota (Prius) and Honda (Insight), but an electric vehicle with an auxiliary gasoline-powered generator – the basic structure of many proposed PHEVs.

Possibly Wouk's title should be changed to "father of the PHEV," the title currently bestowed by some on Andrew Frank, professor of Mechanical and Aeronautical Engineering at the University of California at Davis.[6] According to author Sherry Boschert, Frank had students build a version of a plug-in hybrid for a project when he taught at the University of Wisconsin. In the ensuing 36 years,[7] Frank modified nine gasoline vehicles to run on externally charged batteries and gasoline engines. Whether Wouk or Frank was first or whether one design or another was optimal is not important. What *is* important is that the PHEV concept is 35 years old. Obstacles to success for high-tech products are significant; technological improvement is usually incremental and rarely can be rushed.

The idea of electric and hybrid cars was not limited to universities, research labs and tinkerers' garages. In 1976 the U.S. Congress enacted Public Law 94-413, the Electric and Hybrid Vehicle Research, Development, and Demonstration Act. Among the law's objectives were to work with industry to improve batteries, motors, controllers, and other hybrid-electric components.[8] Thus, the U.S. government has been pursuing electric vehicles for over 30 years. The 1970s may be marked as the decade of the first efforts toward combined electric and gasoline powered automobiles.

The 1990s – Fuel Cells, Electric Vehicles, and Hybrids

There are four core drive train technologies for vehicles that run off electricity. One is the fuel cell electric vehicle (FCEV), whose source of power is hydrogen. The FCEV converts hydrogen to electricity, providing power for an electric motor. The second is the BEV (battery electric vehicle), first invented in the 19th century and produced long before the internal combustion engine (ICE). The third is the HEV (hybrid electric vehicle), best represented by the Toyota

Prius. The fourth is the PHEV (pluggable hybrid electric vehicle). The HEV is essentially an ICE car with a battery assist; the PHEV is a BEV with an ICE assist.

FCEV (Fuel Cell Electric Vehicle) – For decades (since 1970), researchers have proposed an FCEV that would use hydrogen as a fuel, which would emit only water when burned in a process to generate electricity to drive the vehicle. Hydrogen is touted as the most abundant element in the universe (although it is not so on Earth). Ballard Power Systems, a Canadian company, was founded by Geoffrey Ballard, a Canadian geophysicist and businessman, to develop FCEVs. A long-time advocate of replacing the ICE, Ballard founded his first fuel cell company in 1979 to develop commercial applications of the proton exchange membrane fuel cell. He originally began his work in the United States and later transferred it to Canada. He has been acknowledged as the "father of the fuel cell industry." *Time Magazine* named him a "Hero for the Planet" in 1999.[9]

Ballard's first FCEV was a passenger bus delivered in 1993. Interest was generated from this first example and many buses have been built and operated successfully. There was strong interest in this technology from the automobile companies. In 1997 Daimler and Ford invested $750 million for a minority share in Ballard Power Systems, showing a strong commitment to the FCEV concept.[10] In 2001 Ford invested more funds in Ballard and announced a plan to ship the first FCEV in 2004. (This optimism was buttressed by the formation of the FCEV technology-based U.S. Freedom Car consortium in 2002.) Unfortunately, the effort was not successful. Eventually (in 2007), Ford and Chrysler bought the automobile fuel cell technology from Ballard, who then withdrew from the automobile fuel cell engine business.[11] Although the first demonstration was in 1993, and other buses were built, passenger car progress was slow. About one billion dollars was invested in Ballard, mostly by Ford, DaimlerChrysler, and the Canadian government. Ballard's fortunes as measured by stock prices reflect the initial exhilaration and the subsequent decline, increasing from $3.00 a share in 1996 to $129.00 in 2000, to less than a dollar in 2009. This is reflective of the state of the FCEV – an optimistic beginning, a huge peak in interest and investments, followed by a steep decline.

More than a decade was lost pursuing the fuel cell dream. Investment in the FCEV exceeded $15 billion.[12] In spite of a few car companies eventually developing some concept models, no commercial product has come from the multi-billion dollar effort. In a 2008 *Wall Street Journal* article, the two largest car companies in the world, Toyota and GM, expressed their doubt about the FCEV's future.[13] The GM fuel cell models that were made in small numbers cost about $1.5 million each, a higher price than the one quoted five years before.[14]

The complexity of a fuel cell automobile was underestimated. Proponents did not accurately communicate the limitations to the general public. The contribution of Geoffrey Ballard and his team is not to be undervalued. Ballard himself had worked in the energy business and in the 1970s began his research for developing electric alternatives to the internal combustion engine.[15] But his heroic and well-funded effort did not bear fruit, nor have developments since that time provided production models. This should be a cautionary tale about new advanced technology.

BEV (Battery Electric Vehicle) also known as ZEV (Zero Emissions Vehicle) – After Wouk's and Frank's efforts of the 1970s, little new research was conducted on electric cars until GM decided to build a battery electric car (the Electric Vehicle 1 [EV1]) in the early 1990s. The EV1 was based on a prototype vehicle called the GM Impact, created by AeroVironment and introduced at the January 1990 Los Angeles Auto Show.[16 17] Its announcement may have played a part in the passing of a 1990 law in California known as the Zero Emission Vehicle Mandate, an effort to curb California's growing air pollution problem. (Note that the E in ZEV stands for emissions while the E in BEV stands for electric.) This mandate specified that by 1998, 2% of all new cars sold by the seven major auto manufacturers in the state of California were to meet "zero emission" standards as defined by the California Air Resources Board (CARB).

GM shipped the first EV1 in 1997. Eventually, members of the American Automobile Manufacturers Association (plus Toyota, Nissan, and Honda) built approximately 5,600 battery electric vehicles, about 80% sold in California with its moderate climate. GM built 1,100 EV1s and leased 800 of them. Toyota built about 1,575 RAV4 EV electric vehicles, based on the company's conventional RAV4 design. (The RAV4 EV also shipped in 1997.)[18] Toyota found that the RAV4 EV did not sell well even with a $12,000 tax rebate. The company said it made a significant marketing effort but sold only about 300 vehicles per year.

In 2001 CARB modified the ZEV Mandate of 1990 to allow manufacturers

Toyota and Honda to claim partial ZEV credit for the Prius and Insight gasoline hybrid vehicles. GM and DaimlerChrysler then sued the state of California and CARB, alleging the new ZEV rules violated a federal law barring states from regulating fuel economy. CARB removed the requirement for pure ZEV vehicles from the ZEV Mandate in 2003 at which point GM, Toyota, and the other participating manufacturers terminated their BEV programs. (This was the same year that the U.S. government passed the Fuel Cell Initiative to support the Freedom Car.) CARB also began supporting the fuel cell concept as the preferred zero emission option.

The thirteen year period from the first display of the GM Impact in 1990, to the delivery of a ZEV/BEV in 1996, to termination of these models in 2003 advanced the state of battery electric vehicles significantly but not enough to make a competitive and profitable product line. Toyota was able to use valuable technology from the development in the second generation Prius's Hybrid Synergy Drive.[19] [20] GM is using technology from the EV1 program in its Volt development.

GM was attacked for stopping production of the EV1 when CARB changed their rules. A documentary film *Who Killed the Electric Car?,* made in 2006, vilified the company for its supposedly poor decision.[21] However, another view is that the BEV was killed by the success of the first gasoline hybrids (HEVs), the Toyota Prius and Honda Insight, whose development overlapped that of the ZEV/ BEV required by CARB. The Prius shipped in 1997 (the same year as the EV1) and CARB had four years to review the technology as well as measure its market acceptance. This may have led to their 2003 decision to include gasoline hybrids as part of the CARB mandate. This suggests an alternative version, less critical of GM, that Toyota and Honda killed the BEV/ZEV by developing a much better alternative to reduce emissions – the high-mileage HEV.

While GM took back all its EV1s – a condition that was part of the EV1 lease agreement – Toyota left some RAV4 EVs in the hands of customers and was not vilified. The attack on GM was remarkable. At no time did it include a large number of attackers. There are no statistics to determine how many drivers liked or disliked the EV1. Obviously, some were upset that GM took them back. Some probably felt indifferent and some may have disliked the driving experience. Apparently, the tens of millions of California car buyers driving gasoline cars were not nearly as enthusiastic as the few thousands of people who leased electric models.

GM was also heavily criticized for "crushing" the returned EV1s as if there were some malice being displayed. Since all old cars are "crushed" to transport them to recycling centers, this was not a unique negative action taken against the EV1. A more accurate description would be that GM recycled the EV1s. This connotation is less negative but does not serve the purposes of the anti-GM people. More research might find that the campaign was directed not so much to embarrass GM but to justify a position that the BEV has already been proven both technically and in the market, and only the machinations of the car companies limited its success. Whether the end of the BEV of that period was due to price, uncertainty about the car, or poor marketing on the part of manufacturers will probably never be determined. Since the BEV is now being resurrected around a new lithium battery technology, the concept will be retested.

HEV (Hybrid Electric Vehicle: The Diesel Version) – The U.S. government passed the Electric and Hybrid Vehicle Research, Development, and Demonstration Act in 1976. Seventeen years later, the government and the domestic auto industry formed the Partnership for a New Generation of Vehicles (PNGV) to produce family-sized fuel-efficient vehicles that could get 80 mpg.[22][23][24] Involved parties included Department of Energy (DOE) national laboratories, universities, suppliers, and the United States Council for Automotive Research (USCAR) – a legal partnership between DaimlerChrysler Corporation, Ford Motor Company, and General Motors Corporation. GM, Ford, and Chrysler successfully developed high-mileage concept PNGVs. In 2000 GM completed the concept car, Precept,[25] which achieved 80 mpg. Ford's Prodigy was introduced in 2000 and got 72 mpg.[26] Chrysler's ESX-3 also got 72 mpg.[27] All three cars were diesel HEVs. A variety of other innovative technologies were tried in the three vehicles. In a 2001 speech, Al Gore summarized the very successful work done.[28] During the entire term of the Clinton-Gore administration (1993–2001), U.S. companies developed some variations of hybrids and electric vehicles partially funded by the government.[29][30] (Note that this eight-year period overlapped the period of the BEV/ZEV, which began in 1990, first shipped in 1997, and ended in 2003 with the withdrawal of the cars from the market.)

The PNGV partnership was dissolved in 2001 (during the George W. Bush administration) at the request of GM, Ford, and Chrysler. It was replaced the following year when the DOE and USCAR formed a new cooperative automotive research partnership called Freedom CAR, chartered to develop technologies for hydrogen-powered fuel cell electric vehicles.[31] (CAR stands for Cooperative Automobile Research.) In addition to the DOE, partners included oil companies (BP America, Chevron, ConocoPhillips, and Exxon Mobil Corporation), Shell Hydrogen LLC, and the members of USCAR. This represented the combination of the U.S. government, oil companies, and car companies committing to the fuel cell electric car and officially abandoning the diesel hybrid cars that had been under development for eight years. It also effectively cancelled the BEVs/ZEVs initiated by the CARB program in California. The fuel cell Freedom CAR never saw the light of day and the program was redirected to include PHEVs.

HEV (Hybrid Electric Vehicle: The Gasoline Version) – Toyota began its gasoline hybrid effort in September 1993, when it created G21, a committee to research cars for the 21st century. ("G" stands for "globe.") Toyota was excluded from the U.S. PNGV program, which also began in 1993; this may have provided some motivation for the G21 effort.[32] The first G21 meeting was held in February 1994, at which time the team determined that the goal of G21 was to build a car that would be resource and environmentally friendly while retaining the benefits of a modern car. In late 1994 the G21 team designed a concept car with a hybrid gasoline engine for the 1995 Tokyo Motor Show. The concept car was named "Prius," the Latin word for "prior" or "before."

In 1995 the hybrid development was approved by top management and the exclusive hybrid power train was labeled the Toyota Hybrid System (THS). In late 1995 the Prius hybrid concept car made its debut at the 31st Tokyo Motor Show. The car was officially unveiled to the press in October 1997 and went on sale in December 1997. First-year sales were about 18,000 cars. In April 2000 the Sierra Club presented its "Excellence in Environmental Engineering Award" to the Prius. In August 2000 the Prius was launched in the United States as a 2001 model. In March 2001 *Automotive Engineering International*, the official publication of the Society of Automotive Engineers named the Toyota Prius the "Best Engineered Car of 2001." Prius sales topped 100,000 cars worldwide in 2002. Toyota labeled the Prius a Hybrid Synergy Car.[33 34] Toyota was the first car company that committed to the gasoline hybrid design and brought it to market. Together Honda and Toyota have more than 80% of the hybrid cars sold as of mid-2009.

In April 2003 Toyota introduced the second generation 2004 Prius at the

New York Auto Show. Toyota noted that its new, breakthrough electric/gas "Hybrid Synergy Drive" (HSD) system had five key advantages over current hybrids: more power, better mileage, super-ultra-low emissions, more compact, and adaptable to a wider variety of vehicles. Cumulative Prius sales by early 2008 were about 1,000,000 units.[35] A third version of the Prius was developed with shipments beginning in the summer of 2009 in the United States.

Although not as successful as Toyota, Honda also developed high-mileage hybrid gasoline cars. The Honda Insight was introduced into the United States in 1999. The Honda Civic hybrid version was introduced in 2002. The Insight was the highest mileage car in the country but seated only two people and sales were low. It was withdrawn from the market in 2007. In 2009 Honda began selling a new five-passenger car with the same name (Insight), its third small hybrid.

In December 2008 the Honda chief executive officer (CEO) at that time, Takeo Fukui, in his annual year-end speech, noted that more than ever, Honda would concentrate its resources on the development of fuel-efficient products. He said:

> The keys to achieving this goal are advancements of electromotive technologies, including hybrid models as well as motorcycles and small cars. Honda recognizes hybrid technologies as the most realistic path for CO_2 reduction at this moment. To effectively utilize fixed resources, Honda believes it is most important to focus all of our energy on hybrids and achieve mass market penetration as soon as possible.[36]

Honda ranks second in hybrids after Toyota. It announced its latest generation hybrid, the five-passenger Insight, in January 2009. It was priced well below the new Prius. In July 2009 Honda's new president, Takanobu Ito, made an even stronger commitment to hybrids,[37] saying that the company would speed up its hybrid vehicle plans and offer a CR-Z sporty hybrid model in Japan in February 2010. A hybrid version of the Fit subcompact will be available before the end of 2010. Ito said hybrids are so crucial to the industry that over the next 20 years every model, including those from other automakers, will be a hybrid.

Reviewing 1990 to Today – Of the four core technologies of this period, the three U.S. ones failed and the Japanese one succeeded. The U.S. BEV effort covered the period from 1990 to 2003 (14 years); there are no more than a few hundred vehicles from this program still running. The FCEV period can be measured from the work of Ballard, which began in the late 1980s, to the sale of its technology in 2007 (about 20 years). The diesel HEV development period (U.S. PNGV program) ran from 1993 to 2001 (eight years). The successful gasoline HEV began in 1993 and continues in its original form today, with consistently well-thought-out improvements from Toyota and Honda.

Detroit's Era of Woe – Losing 15 Years

The decades of the 1990s and 2000s marked a special period in automobile history that resulted in Toyota becoming the world's largest manufacturer while GM and Chrysler filed for bankruptcy. The BEV/ZEV, FCEV, and diesel HEV were major failures that consumed significant resources of American manufacturers. The results are shown in Figure 1. Toyota is winning both the technology and marketing race.[38]

Figure 1: Global Vehicle Production

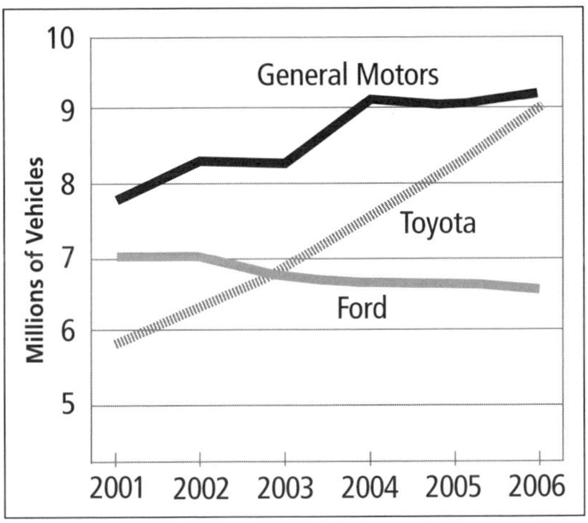

The U.S. car companies have lost their historically dominant position and other countries now match or exceed U.S. research and development (R&D) investments. Table 1 shows the breakdown of R&D of the major car companies as well as revenue projected from the R&D spending. R&D as a percent of 2007 sales is also shown.

The financial success of the top four corporations can be compared by the total values of the company's stock. In early 2009, Toyota's worth was $104 billion, Honda $79 billion, GM $2.2 billion, and Ford $5.88 billion.[39]

In terms of gas mileage, the record of the U.S. manufacturers is poor. Table 2 shows the top ten fuel-efficient cars.[40]

Every region of the globe is increasing its automobile sales while the U.S. share shrinks. (See Figure 2.)

Two of the top three U.S. companies are now more or less owned by the government. As the companies begin to make a comeback, unfortunately, they are falling prey to yet another high-risk untested techno-fix: the PHEV.

Table 1: R&D and Percent of Sales

Company	2007 R&D in billions	% of 2007 Sales	Revenue in billions
Toyota	$8.39	3.6	$233.1
GM	$8.10	4.5	$180.0
Ford	$7.50	4.3	$174.4
Honda	$5.14	4.9	$104.9
Volkswagen	$4.76	3.2	$148.7
Daimler	$4.32	3.2	$135.0
Nissan	$4.00	4.2	$95.2
BMW	$4.00	5.2	$76.9
Peugeot	$2.84	3.4	$83.5
Densco	$2.72	7.7	$35.3

Table 2

	Automobile Model	MPG city/hwy
1	Toyota Prius Hybrid	48/45
2	Honda Civic Hybrid	40/45
3	Nissan Altima Hybrid	35/33
4	Ford Escape Hybrid FWD	
	Mazda Tribute Hybrid 2WD	
	Mercury Mariner Hybrid FWD	34/31
5	Smart Fortwo Convertible	
	Smart Fortwo Convertible	
	Smart Fortwo Coupe	33/41
6	Toyota Camry Hybrid	33/34
7	Volkswagen Jetta (manual, diesel)	
	Volkswagen Jetta Sportwagon (manual, diesel)	30/41
8	Volkswagen Jetta (automatic, diesel)	
	Volkswagen Jetta Sportwagon (automatic, diesel)	29/40
9	Toyota Yaris (manual)	29/36
10	Toyota Yaris (automatic)	29/35

Creating a PHEV Brand – The Role of EPRI

In the past, car companies, oil companies, and various U.S. governmental agencies cooperated in pushing the private car over mass transportation options. It should be no surprise to discover a relationship between coal companies and electric utilities to push for electric cars to replace gasoline vehicles. One of the

Figure 2: Number of Vehicles Manufactured by Country

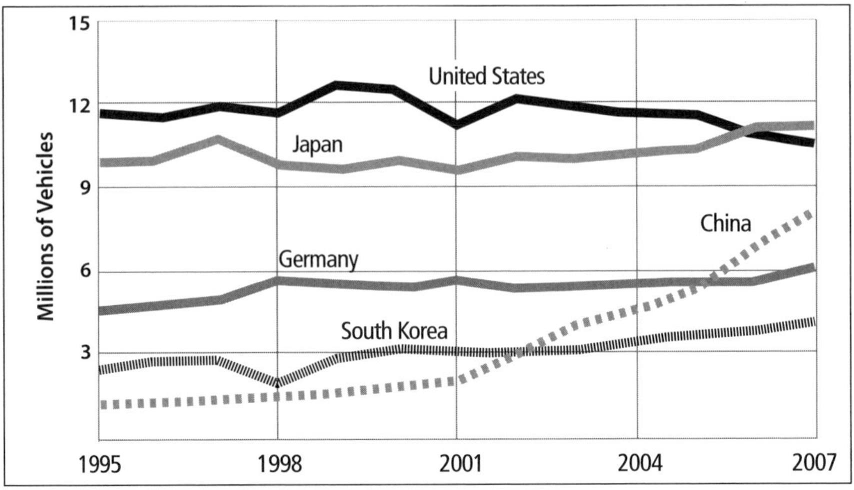

key organizations supporting this direction is the Electric Power Research Institute (EPRI). EPRI is a nonprofit organization, funded by the nation's electrical utilities and established in 1973 to advance the technology of electricity production. As the American Petroleum Institute is the main U.S. trade organization for the oil and natural gas industry, so EPRI is the key trade organization for electrical utilities. EPRI conducts research and advocates for its positions with local, regional, and national organizations, including car manufacturers and government agencies. EPRI's stated purpose[41] is as follows:

> The Electric Power Research Institute (EPRI) leads research, development, and demonstration of technical and operational solutions in electricity generation, delivery, and use. The focus and application of EPRI's research and activities span virtually every aspect of the power industry, including reliability, safety, the environment, and energy efficiency. The Institute's collaborative model engages EPRI members, participants, scientists, and engineers, along with experts from academia and other business sectors. As an independent, nonprofit center for public-interest energy and environmental research, EPRI's work is supported both by its members, which represent more than 90% of the electricity generated in the United States, and by growing international participation, representing more than 15% of EPRI's program support.

EPRI has lobbied for electrical vehicles for some time. The Institute's electric vehicle program began in 1976. In 1991 EPRI joined with GM, Ford, Chrysler, and the Department of Energy (DOE) to found the U.S. Advanced Battery

Consortium (USABC). USABC's purpose is to advance electric and hybrid vehicle battery technologies. In 1999 EPRI formed the Hybrid Electric Vehicle Working Group with utilities including Southern California Edison, the New York Power Authority, and Southern Company, along with members of the automotive industry, the U.S. government, regulatory agencies, and university research organizations. This consortium focuses on studies and analyses of PHEV technology and market acceptance. In 2003 EPRI developed a partnership with DaimlerChrysler to design, develop, test, and demonstrate the PHEV Sprinter, a Mercedes-Benz light commercial vehicle prototype, at various sites in the United States.

EPRI has collaborated with government organizations, auto manufacturers, and electric utilities in the development, testing, and demonstration of electric-fueled transportation technologies.[42] It advocates for development of PHEV technology as an alternative to petroleum-based transportation. EPRI defined its role and the early PHEV concepts in a key 2001 paper entitled "Comparing the Benefits and Impacts of Hybrid Electric Vehicle Options."[43] Organizations that participated in the preparation of that report included Arthur D. Little, Inc., Southern California Edison, EPRI, the University of California Davis Hybrid Electric Vehicle Center, GM, Ford, Argonne National Laboratory, National Renewable Energy Laboratory (NREL), Applied Decision Analysis, the DOE, and Sacramento Municipal Utility District. Key individuals who participated included Fritz Kalhammer and Robert Graham (both with EPRI at the time) as well as Andrew Frank and Mark Duvall from the University of California at Davis. (Duvall later joined EPRI and heads up its PHEV effort.) This document laid out the rationale for PHEV and BEV vehicles. It defined different PHEVs based on their distance driven on batteries as well as their configuration as serial or parallel. Note that 2001 was the same year that GM, Ford, and Chrysler requested that the PNGV be canceled and a year before the fuel-cell-focused Freedom CAR program was initiated.

EPRI continues to advocate for electric vehicles.[44] A huge business opportunity exists for utilities if America's 230 million cars are replaced with electric vehicles. EPRI's support of the PHEV marketing program is a logical step for substantially increasing revenues of its supporting utility companies. No PHEV car is being developed by EPRI; its role is to excite the public and pressure the auto industries to develop such cars.

Starting a Movement – Helping EPRI

Various private groups have been formed to support the PHEV concept. CalCars (also known as the California Cars Initiative) is a nonprofit organization founded in 2002 by Felix Kramer, a provider of communications, business development, and marketing services for startup companies. He intended to promote what he calls the clean, advanced technology of PHEVs as a key to addressing global warming.[45] CalCars focuses on advocacy to prove market demand and on technology development to demonstrate that PHEVs can be extensions of current model lines. The organization's stated intention is to force car makers to build PHEVs, by applying public pressure and enacting government legislation. Kramer works closely with Mark Duvall, director of electric transportation at EPRI.

In September 2004 Kramer and some California engineers converted a 2004 Toyota Prius into a PHEV, which they labeled PRIUS+. This was achieved by adding 300 pounds of lead acid batteries to the car and connecting these batteries to the electric motor of the Prius. The CalCars website says that the PRIUS+ achieved roughly double the fuel economy of a standard Prius and could make trips of up to nine miles using only electric power. Maximum speed was 34 miles per hour for the modified car when operating under battery power (a detail rarely mentioned in the press).

In May 2006 CalCars flew a Prius version with lithium ion batteries to Washington, D.C. Numerous governors, senators, and members of the House of Representatives rode around the capital in the modified Prius. There was tremendous enthusiasm generated by the PHEV presentations. During this trip Kramer and Duvall, together with Andrew Frank of UC Davis, met with members of Congress and spoke to congressional committees. Senators expressed their amazement at the modified Toyota Prius. Its proponents were extolled as visionary geniuses who, in a backyard, had developed products that the global car manufacturers could not. The implication was that a few bright people had done to a Prius what Toyota had not been able to do or had overlooked.

What made the car such a resounding success was the banner painted on the models stating "100+ miles per gallon," a theme on the websites and literature of PHEV advocates. (The 34 mile per hour [mph] maximum speed is not part of the label on the car or emphasized in the literature.) Congressional members were viewing a car that supposedly got almost five times the average 21 mpg of the existing U.S. car fleet. And it could be done for existing cars at a cost of a few thousand dollars! The constantly repeated messages of "the technology is here" or "just around the corner" or "it's proven" was reassuring even though inaccurate. By the end of this public relations campaign, both presidential candidates were committed to high-tech PHEVs. The PHEV marketing campaign of CalCars, EPRI, and others was successful in reaching top politicians. The information they presented was effective. Whether it was accurate is another story.

PHEV Brand Creation – Political and Technical Supporters

It was not simply CalCars and EPRI that influenced the presidential candidates. Senator Joe Lieberman in a Loewy Lecture at Georgetown University on October 7, 2005 said, "And then we must remake our automobile engines as well. Vehicles that get 500 miles per gallon – or that use *no* refined crude oil – are within our grasp." He further noted,

> But we can do even better – dramatically better – with the plug-in hybrid that is just now on the threshold of commercialization. Like the present hybrids, it would use both a gasoline and electric motor. But the plug-in hybrid would be able to use the battery exclusively for the first 30 miles of a trip. That means a plug-in hybrid would use zero – zero – gallons of gas or any combustible fuel for the vast majority of its trips. And experts tell me it could effectively get the 500 miles per gallon on longer trips. This isn't pie in the sky. These vehicles could be in your garage within a couple of years.[46]

Note that the "couple of years" quoted would have been 2007 from the date of his speech. In 2009 there were no such vehicles except for a few experimental models.

Lieberman's speech included the phrase "experts tell me that..." Who are some of the experts responsible for the 500 mpg quote? Two are R. James Woolsey, ex-CIA director, and George Schultz (ex-Secretary of Labor, the Treasury, and State), co-chairs of the Committee on the Present Danger. (Joseph Lieberman is an honorary co-chair of this organization.) In a 2005 policy paper entitled "Oil and Security," Woolsey and Schultz developed the 500 mpg claim.[47] Author Sherry Boschert, in her book *Plug-in Hybrids: The Cars That Will Recharge America,* quotes from this paper:

> A hybrid that gets 50 miles per gallon could get 100 miles per gallon if made from carbon composites that would be lighter but no more expensive than

> conventional materials, they wrote. Run that lightweight hybrid on 85% cellulosic ethanol or a similar proportion of biodiesel, and it would get hundreds of miles per gallon of gasoline or petroleum derived fuel. Upgrade that car to a plug-in hybrid that runs 30 miles on electricity before tapping the liquid fuel and it might get 1,000 miles per gallon of gasoline.[48]

In a *Wall Street Journal* article entitled "Gentlemen, Start Your Plug-ins," Woolsey states, "A 50-mpg hybrid, once it becomes plug-in, will likely get over 100 mpg of gasoline (call it "mpgg"); if it is also a flexible fuel vehicle using 85% ethanol (E-85), its mpgg rises to around 500."[49] Later Woolsey penned another *Wall Street Journal* article with Paula Dobriansky noting that PHEVs would get more than 100 mpg and 500 mpg with E85.[50]

Another well-known expert is Joseph Romm, author of the 2004 book *The Hype About Hydrogen: Fact and Fiction in the Race to Save the Climate.* Romm and Andrew Frank (the so-called "father of the PHEV") co-authored a paper entitled "Hybrid Vehicles Gain Traction,"[51] which included the following statement: "These kinds of vehicles could travel 500 miles on one gallon of gasoline blended with five gallons of ethanol and thus continue a long-term strategy for dealing with the inevitable peak and subsequent decline in world oil supplies." Romm commented further on the PHEV in a paper he and Peter Fox-Penner jointly wrote in *Electric Wheels,* with the title "America needs to reduce its gasoline consumption. But how? By promoting plug-in hybrid electric vehicles."[52] The authors refer to the PHEV as one emerging technology with the potential to trigger transformative change in just five to ten years if government takes a few smart steps to spur its commercialization. They note,

> In all-around driving, plug-in hybrids could average between 80 miles per gallon and 160 mpg, compared with about 45 mpg for the best-known of today's conventional hybrids, the Toyota Prius. The gasoline savings could be even greater if plug-in hybrids were designed to run on biofuels; they could travel 500 miles on a gallon of gasoline blended with five gallons of ethanol.

One utility took a significant role in the branding of the PHEV effort. Austin Energy Company provides electricity to the capital city of Texas. It is a municipally owned electric utility that serves 388,000 residential and commercial customers with 2,600 megawatts (MW) of generation capacity primarily from its interests in three gas-fired plants, a coal-fueled facility, and a 16% ownership of a nuclear-powered electrical generating plant.[53] Renewable energy sources include wind power (214 MW) and landfill methane (13

MW). Power deregulation took effect in Texas in 2002 but as a city-owned utility, Austin Energy was not required to participate.

In January 2006 Roger Duncan, general manager of Austin Energy, formed Plug-In Partners to advocate for the PHEV by influencing other cities and garnering support to press manufacturers to build the cars. Part of their program was to get what they called "soft orders," essentially small refundable deposits for mythical PHEVs, which were to show manufacturers that high public interest existed. The program was ended in October, 2008.[54] The results, as reported on the Plug-In Partners website, showed that, "The Plug-In Partners effort over the last three years that enlisted 631 partners, signed up 38,312 petitions and made 11,333 soft purchase orders and participated with outreach and educational efforts in over 88 conferences and expositions has been a critical success."[55] The soft purchase orders showed the auto manufacturers that after three years of marketing, fewer than 12,000 out of hundreds of millions of drivers were willing to make a refundable deposit. Partners and petitions (whatever these may represent) also show a small number of involved people, hardly justifying the claim for a national movement. Nonetheless, this was heralded as a major triumph with Austin Energy stating that they had accomplished their mission and were now ending the program.

Plug In America is a nonprofit organization that also advocates for PHEVs. In a presentation to the California Air Resources Board in 2006, Plug In America stated that "Plug-in vehicles are practical, proven, and ready," noting that consumer demand is already demonstrated (presumably by the few hundreds of people who are driving early versions of no longer marketed battery vehicles).[56] The same presentation notes that 150 million emission-free *consumer* miles have already been driven. This number was derived not from PHEVs but from the BEVs made as a result of the CARB program. One hundred fifty million miles have certainly not been driven in PHEVs! This is symptomatic of the overstatement of the PHEV case. In 2008 Plug In America received their 501c3 designation and hired their first employee.[57]

Deconstructing the False Mileage Claims

The current PHEVs would be an interesting laboratory curiosity except for one thing – the claims of high gas mileage from the President of the United States on down. The quoting of the 150 mpg number is an example of "puffery."

Puffery is a legal term that refers to promotional statements and claims that express subjective rather than objective views, such that no reasonable person would take them literally. The principle was confirmed in law that certain statements made by advertisers that were obviously not made in a serious manner could be exempt from usual rules relating to promises in open contracts. The U.S. Federal Trade Commission defined puffery as a "term frequently used to denote the exaggerations reasonably to be expected of a seller as to the degree of quality of his product, the truth or falsity of which cannot be precisely determined." Puffery is not illegal in this country. In fact, it is endemic to our business and personal relationships. CalCars, EPRI, Austin Energy, and Plug In America, as well as individuals such as Romm, Woolsey, Schultz, Kramer, Frank, and Duvall have been extremely successful with the use of puffery in their PHEV campaign. Their success is based on the idea of a "proven" technology (a few dozen backyard modifications) that gets incredible mileage – 100 to 150 mpg. There have been no test cases and no analysis of exactly what the claims mean. But we live in a "pufferous" culture so being mislead comes with the territory. Three examples exemplify the approach.

CalCars – Although not a manufacturer, CalCars has a strong effect on the positive perception of the PHEV. Their story is simple – the gas mileage of a modified Prius is doubled. Thus, their cars include the note "100+ mpg" painted on the side. Their modes of calculation have not been published but can be derived, as will be shown later in this document.

AFS Version of PHEV – Hype is not limited to the Prius versions. In November 2008 AFS Trinity Power Corporation, a Bellevue, Washington company, withdrew their plug-in hybrid (a modified Saturn Vue SUV) from the Los Angeles Auto Show after the show management refused to allow them to advertise their product as a 150+ mpg car. AFS had been showing the car for about a year, taking it to Washington in April of 2008, duplicating CalCars similar effort in 2006 to lobby Congress but with a different vehicle. AFS implied that the position of the auto show management was based on jealousy or the wish to hide AFS's technical advantages. AFS parked their display and the modified cars on the roadway near the convention and gave test rides and literature.[58]

AFS explained their position in a press release pointing out that the show sponsors wanted the companies' mileage claims verified by the EPA, which measures mileage standards.[59] The report noted,

> AFS Trinity's 150-MPG mileage estimate is based on a typical week of driving: 40 miles, 6 days per week and 80 miles on one weekend day each week. The first 40 miles of every day are electric, with gasoline being used only for distances beyond 40 miles. For this driving profile, the XH-150 uses up to 2 gallons of gas to travel 320 miles, which works out to 160 mpg, which the company rounds down to 150 miles per gallon.

This is similar to the CalCars method of calculation for the modified Prius.

The AFS Trinity mileage calculation process is shown in a graphic (Figure 3) on their website.[60] AFS assumes a distribution of 80% electricity (running off the batteries) and 20% normal mode (running off the engine). The modified basic car normally gets 30 mpg. In this example, AFS quotes 150 mpg based on dividing the 30 mpg of the gasoline engine by 20%, the portion driven using the gasoline engine.

Figure 3

This is not a concept car. It is running now.

80% Electricity + 20% Hybrid = >150 MPG*

(∞ MPG) *(30 MPG*)*

GM Volt – GM vice-chairman Bob Lutz introduced the Volt concept at the 2007 North American International Auto Show[61] as a "battery-powered, four-passenger electric vehicle that uses a gas engine to create additional electricity to extend its range." (This is not a bad definition of a PHEV.) He also noted, "If you lived within 30 miles from work (60 miles round trip) and charged your vehicle every night when you came home or during the day at work, you would get 150 miles per gallon." There are four specifications of the Volt that are used to calculate mpg:

Miles per gallon using gasoline generator is 50 mpg.

The car has a 12-gallon gasoline tank.

The car can travel 40 miles on a fully charged battery.

Distance drivable on one tank is 640 miles.

(This is used to verify the statistics: 640 miles equals 50 mpg x 12 gallons for 600 miles plus the 40 miles from the charged battery.)

Doing the Math – and Exposing Duplicity

False Mileage Claims – Option 1 – The miles per gallon determination popularly used by PHEV advocates is simple – decide what percent of the distance driven uses the gasoline engine and divide that into the miles per gallon of gas obtained when using the gasoline engine. In the case of CalCars, to get 100+ mpg, they may decide to assume the Prius is driving 40% of the distance using gasoline and divide the 45 mpg of the car by 40% getting 101 mpg. For AFS the assumed percent of distance driven with electricity is 80%. Dividing the 30 mpg gasoline mileage by the 20% distance using the gasoline engine results in 150 mpg.

How does the Volt achieve 150 mpg? Lutz's statement defines the daily trip as 60 miles (the time between charges). The car can go 40 miles on the battery. This leaves 20 miles driven using the gasoline engine, which is one-third of the total trip. Dividing the 50 mpg rating of the gasoline engine by one-third gives 150 mpg.

All the arguments are based on a simple formula that is based on a simple assumption that it makes no difference what the source of electricity for the car may be. The formula divides the miles per gallon obtained when using the gasoline engine by the percent of trip distance using the gasoline engine to calculate the PHEV mpg.

PHEVmpg = ICEmpg divided by the % DistanceDrivenICE

This equation is the basis for mpg calculations. The kind of fuel for generating electricity is unknown. Obviously, a great deal of information is missing that would give a very different picture. The equation does not account for the fossil fuels used to provide the electricity. Such calculations, simple though they may be, are rarely reported. The President of the nation, CEOs of auto companies, heads of utilities, and directors of coal companies do not want to use a more complex equation.

It is useful to run a thought experiment that may clarify this misrepresentation. Suppose a PHEV runs half the time on gasoline and half the time on electricity generated from a gasoline-fueled electricity generator in the owner's garage. Assume the PHEV gets 50 mpg when using the gasoline engine. Also assume that it takes exactly one gallon of gasoline to generate sufficient electricity to charge the batteries enough so that the Prius can drive 50 miles on the charge from the battery. To go 100 miles, half on battery power and half on engine power, requires that one gallon of gasoline is put into the gasoline tank and one gallon of gasoline is put into the tank of the generator used to charge the car's battery. The car would go 50 miles on the charge in the battery and then 50 more miles on the gallon in the gasoline tank. What is the mpg of the PHEV? Recall that two gallons of gas were used to go 100 miles. Thus, the mpg would be 50 miles per gallon. But according to the methods used by CalCars, Trinity, and GM, the car would get 100 mpg. It is tragic that such mathematical chicanery is allowed to confuse the populace (and the President).

False Mileage Claims – Option 2 – The EPA has not been helpful in keeping the public informed about the nature of mileage claims. The agency was providing very high mileage numbers for the BEVs of the late 1990s, which were discussed earlier. For example, the EPA gave the 2003 Toyota RAV4 EV, a battery electric car, a rating of 112 mpg.[62] The EPA quote includes the statement, "For electric vehicles, mpg represents the number of miles that can be traveled using an amount of electricity equivalent to the energy in a gallon of gasoline." This is a familiar misrepresentation. Many neophytes convert the number of BTUs in a gallon of gasoline to an equivalent number of kilowatt hours (kWh). Next they look at the distance electric cars can drive on the kWh equivalent in a gallon of gasoline. This provides a major improvement: in the case of the RAV4 EV, a factor of three. However, this completely ignores the approximately 67% energy cost of converting fossil fuels such as coal to electricity.

At times the EPA and others have made a comparison based on the *cost* of gasoline rather than the energy equivalent. One calculation is given in an example on Wikipedia,[63] which states,

> As of May 2006, charging a RAV4 EV from full-dead to full-charge, at a rate of USD [$] 0.09 per kilowatt-hour, costs around $2.70. As of May 2008, based on a gasoline price-per-gallon cost of $3.80 and up and the non-EV 2003 RAV4 2-wheel-drive gasoline fuel efficiency of 27 $mpg_{\text{-US}}$ (8.7 L/100 km; 32 $mpg_{\text{-imp}}$), the RAV4 EV costs approximately 25% as much to fully charge, and makes mileage in the RAV4 EV the cost equivalent to a 111.1 $mpg_{\text{-US}}$ (2.117 L/100 km; 133.4 $mpg_{\text{-imp}}$) small SUV.

This results in the equivalent mileage based on some assumed costs. Unfortunately, this continues to mislead the public, which expects the miles per gallon to be actual miles traveled per quantity of fuel put into the vehicle. Using the preceding approach, the miles per gallon would change, depending on the price of gasoline or electricity.

Doing the Math – and Exposing More Duplicity

The duplicity of car companies and their supporters is endless. And government groups such as EPA and DOE, which have supported the car companies' R&D programs for decades, typically play important roles in misleading the consumer. In August 2009, GM president, Fritz Henderson, made a major announcement as part of a marketing campaign extolling the miles per gallon for the upcoming Volt PHEV. The statistic of 230 mpg was given and reference was made to a new EPA rating for PHEVs and electric cars. The EPA shortly thereafter issued the following statement:

> EPA has not tested a Chevy Volt and therefore cannot confirm the fuel economy values claimed by GM. EPA does applaud GM's commitment to designing and building the car of the future – an American-made car that will save families money, significantly reduce our dependence on foreign oil and create good-paying American jobs. We're proud to see American companies and American workers leading the world in the clean energy innovations that will shape the 21st century economy.

When contacted for comment, GM said that EPA is not backing away from the 230 mpg statistic and that it's unlikely that EPA will come out with a much lower number when they actually test a Volt. Furthermore, Henderson noted that GM believes that coming out with the 230 mpg rating at this time is one way to change people's perception of "what kind of car the Volt is."[64]

The fact that EPA did not immediately challenge the number shows the government support for GM. With the United States owning most of the corporation, it now finds itself supporting the puffery extolled by the company's ranking chief officer. As far as the idea of telling people what kind of car the Volt is, the charade really tells people what kind of company GM is, as well as the kind of government we have. This will be clear later when CO_2 comparisons are made.

Understanding Power Plants

A "coal car" (or electric car) is somewhat more efficient than a gasoline car because the efficiency of power plants that create the electricity is somewhat higher than that of an internal combustion engine. Although the exact ratios require more detailed analysis, a first approximation might be 30% efficiency for a coal plant (70% lost to heat and transmission) and 20% for a gasoline engine.[65] That means an electric car running on electricity from a power plant would effectively apply 30% of the 10,500 BTUs in a pound of coal (3,150 BTUs). A gasoline car would effectively apply 20% of the 20,000 BTUs in a pound of gasoline (4,000 BTUs). A different grade of coal would make the electric car more efficient. There are no major differences in gasoline so the gasoline car would stay the same. Figure 4 illustrates the high energy losses in converting fossil fuels to electricity.[66]

Figure 4: Energy Sources and Uses

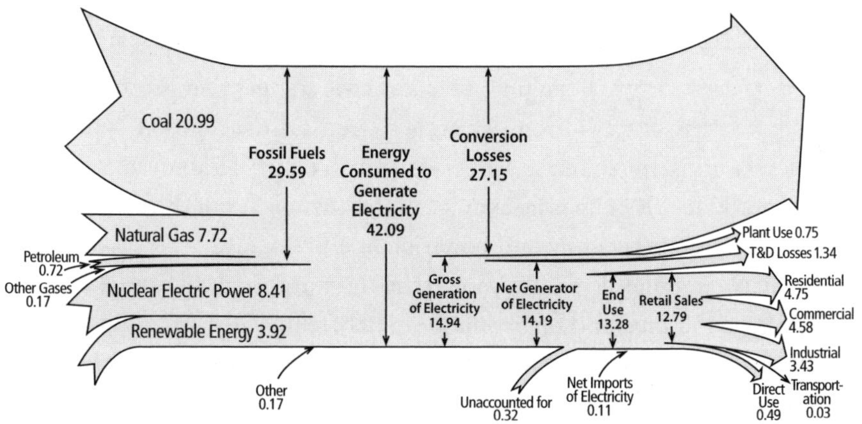

Efficiency of power plants has improved over time but no more rapidly than the efficiency of the internal combustion engine. The evolution of net electric efficiency of natural gas and coal plants on a worldwide basis is shown in Table 3.[67] New Integrated Gasification Combined Cycle (IGCC) plants have offered some improvement but they are still in the early implementation phase.

Table 3

Year	1974	1990	2003
Natural Gas Efficiency	36%	35%	42%
Hard Coal Efficiency	30%	34%	35%

The efficiency of a coal-driven electric car is about the same as that of the gasoline car. But coal provides only about half of the electricity in the country. Natural gas is more efficient. It contains about 22,000 BTUs per pound. The electricity from natural gas available (after accounting for its conversion to electricity) would be about 7,700 BTUs per pound of natural gas, much higher than the 4,000 BTUs in a pound of gasoline. When comparing on a national basis, we would not average the coal and natural gas but weigh it according to its usage, with coal providing a bit more than twice as much electricity as natural gas. Coal, natural gas, oil, and wood account for over 70% of the electricity generated. This is shown in Table 4, which displays the 2009 U.S. fuel mix.[68] (Total yearly electricity generation is about 3,576 billion kWh.)[69]

Like coal, all of these sources lose some energy in transmission and the transmission of electricity, no matter what the fuel is, must be considered. The loss is dependent on the distance from the power plant to the place of use. This may be up to 10% of the net electricity.[70] Diesel engines are more efficient than gasoline engines. Comparisons of the efficiency of gasoline/diesel engines versus electric engines that get their energy from a nation's electrical distribution system are complex, but this simple analysis provides a rough range of improvement – possibly from 50 to 100% in efficiency. And the analysis is not difficult to work out for the nation's fuel mix and power plant efficiency.

Table 4

Coal	44.7%
Natural Gas	23.3%
Nuclear	20.2%
Hydroelectric	6.9%
Other sources	3.9%
Petroleum	1.0%

If this were done, it would show that the purported advantages of the PHEV, e.g., 150 mpg and up, would be greatly tempered. We might develop a different measure, such as miles per gallon equivalent, that would then take into consideration the mix. This would be far better than the misleading way it is currently being done. And it may be equally effective to measure fuels by weight. Or we could simply measure CO_2 generated.

The 230 mpg claim of GM and the obsequious supportive response by EPA have led to a storm of criticisms concerning the hyped PHEV claims. Other agencies of the government are beginning to attempt scientific or science-based comparisons, possibly concerned by auto manufacturers' statements and possibly even the optimistic statements of the President.[71] Researchers from the DOE's National Renewable Energy Laboratory (NREL), Idaho National Laboratory, and Argonne National Laboratory are cooperatively developing a method for predicting the real-world fuel and electricity consumption of PHEVs. According to data on the Hymotion Prius PHEVs, the only PHEV currently available in large numbers, the method shows promise for reasonably predicting PHEV average fuel and electricity use, despite differences in design. NREL's method estimated fuel consumption of 4.2 L/100 km (55 mpg) and electricity consumption of 5.5 kWh/100 km (89 kWh/mi). This represents an average decrease in gasoline fuel consumption of roughly 20% relative to the fuel consumption of a comparable HEV Prius. This is far from the 80% reduction implied by a 230 mpg claim (230 vs. 46 mpg).

CO_2 Comparisons

A change from an engine that uses gasoline from oil as its fuel to an engine that uses electricity from coal power plants, natural gas power plants, or uranium-driven nuclear power plants is merely a change in the source of CO_2 for personal transportation, not an elimination of the CO_2. A PHEV should be evaluated principally in terms of CO_2 emissions. The politics of the Organization of Petroleum Exporting Countries (OPEC), cheaper fuel costs, energy independence and security, or other environmental considerations should be secondary to life-threatening amounts of carbon dioxide. The relationship between energy and CO_2 is not adequately addressed by many supporters of PHEVs, allowing them to make inflated mpg claims. For some the main objective is to stop importing oil from the Middle East. In this case, any reduction in oil use is valuable regardless of the substitutes – as long as there are substitutes. Biofuels, synthetics from coal or natural gas, and liquid fuels from tar sands aid in achieving this objective. But for most knowledgeable people, the key objective is to reduce CO_2 emissions. A warming world has far more serious consequences than paying high prices to OPEC.

Much work has been done on evaluating energy in terms of the CO_2 generated. Moving from the mpg orientation to comparing CO_2 generation is a more useful approach. Several studies have estimated CO_2 emissions from PHEVs compared to gasoline cars to determine reductions in total CO_2 emissions. One such study was made by EPRI and reported in an article from the organization's Spring 2008 *EPRI Journal*.[72] The projection of greenhouse gas savings is displayed in a chart from that report (see Figure 5).

Five hundred million metric tons seems impressive but is less so when considering the portion of the total CO_2 generated. The average American generates about 19 metric tons of CO_2 per year.[73] The EPRI analysis suggests that PHEVs could reduce the consumption by "as much as" (implying an ambitious goal) 500 million metric tons per year which, on a per capita basis, is approximately 1.6 tons per year per person. This is about 8% of the current U.S. annual per capita generation of 19 metric tons.

EPRI's analysis was done in 2008, so the cumulative reductions would be over the 42 years from 2008 to 2050, which would be an annual reduction of about 80 million to 245 million metric tons per year. Averaging these two

Figure 5: Annual Greenhouse Gas Emissions Reductions from PHEV Adoption

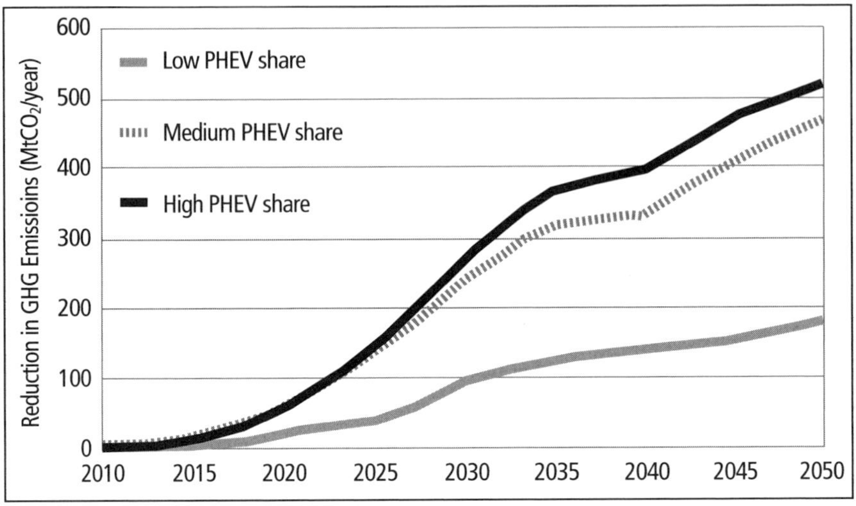

numbers gives 113 million metric tons per year. This is approximately .4 metric tons per capita, which is about 2% of each person's annual emissions, a relatively trivial amount. To meet the Intergovernmental Panel on Climate Change requirements, the world average CO_2 emissions in 2050 will have to be about 1 metric ton per person. For Americans this represents about a 90% cut by 2050. The accumulative savings of .4 metric tons per year up to 2050 is tiny compared to the need. And if we reach that ambitious goal, it still saves us only 1.6 metric tons per year out of the current 19 metric tons per year, an order of magnitude less than what is needed. Shifting from oil for gasoline to coal and natural gas for electricity, even if more efficient, does not really help address the serious climate change issue.

Other EPRI reports show similar results. Two such reports illustrate the evolution of EPRI's thinking about PHEV emissions – one from 2001[74] (see Figure 6) and a second prepared in 2007 (see Figure 7).[75]

In the 2001 report (Figure 6), EPRI showed the advantages of the HEV 0 (a standard HEV) over conventional vehicles. The HEV reduced emissions by about 30%. Note that this does not imply the use of any electricity from an outside source; it is an improved design of a gasoline engine. The HEV 20 and HEV 60 are PHEVs. (This was an early nomenclature for PHEVs.) In this report EPRI shows the HEV 20 (or PHEV) as reducing the emissions about 20–25% compared to a standard HEV.

Figure 6: CO_2 Emissions by Vehicle Type

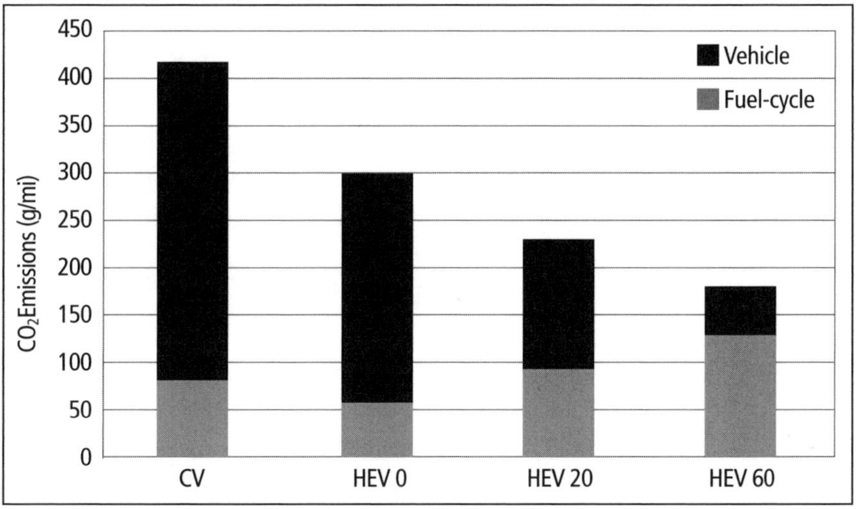

Figure 7: Wells to Wheels Greenhouse Gas Emissions

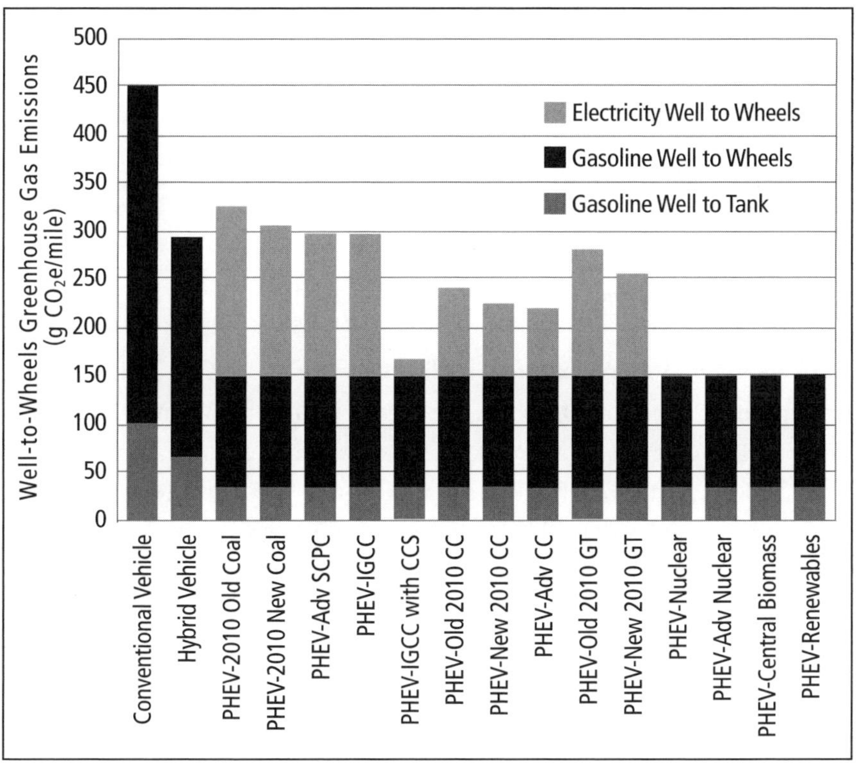

In the 2007 report (Figure 7), the HEV has an advantage over a PHEV 20, at least with the present mix of power plant technologies. More improvement is made with other technologies but each of these carries an associated risk. This is not surprising as new technology efforts are typically optimistic and as the realities of development occur, many theoretical advantages are tempered. The important point is that Figures 6 and 7 show that PHEVs have little advantage over HEVs, at least in the early stages. The PHEV, still largely theoretical, and the HEV, well advanced in the market, are similar in CO_2 emissions. Note that this comparison has a very different connotation from the optimistic views of PHEVs

Figure 8: Change in Cumulative CO_2 Emissions

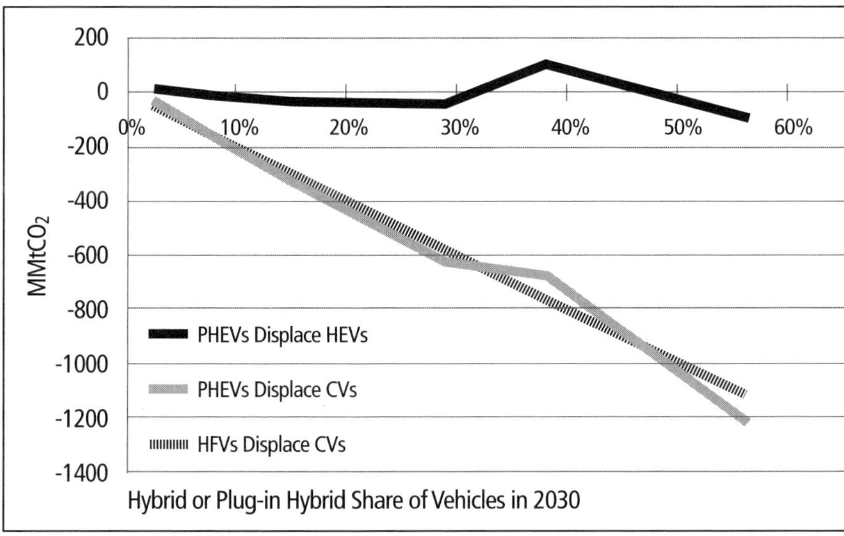

getting 150 or 240 mpg, with an implied large savings in CO_2. Other writers have made the same point as shown in Figure 8.[76] The PHEV is not a significant improvement over the HEV and it has yet to be delivered in quantity!

A 2006 report from the American Council for an Energy-Efficiency Economy concludes that PHEVs will reduce CO_2 emissions somewhat, but not to a large extent.[77] Under the average U.S. power generation mix, the CO_2 reduction of the plug-in relative to a hybrid would be about 15%. Considering the stage of the PHEV at the time that was written (when none were in production), this number might be optimistic. The report also warns about the increase in the number of coal plants to provide PHEV electricity.

> Running on the average U.S. power generation mix, the same plug-in would reduce CO_2 emissions by about 15%. Plug-ins' greenhouse gas [GHG] benefits would grow only in a future scenario in which power sector CO_2 emissions are reduced from present levels. While one might view a dramatic reduction in GHG emissions from power generation as inevitable, given the urgent need to address global warming, the U.S. is experiencing a resurgence in the construction of coal-fired power plants, which could in fact raise the electricity sector's carbon emissions substantially in the coming decades.

The Natural Resources Defense Council (NRDC) also takes a similar position. Figure 9 is from an NRDC report on comparing PHEVs to mid-sized cars.[78] It shows similar comparisons for conventional vehicles, HEVs, and PHEVs.

Figure 9: Global Warming Emissions, PHEVs Compared to Other Mid-Sized Cars

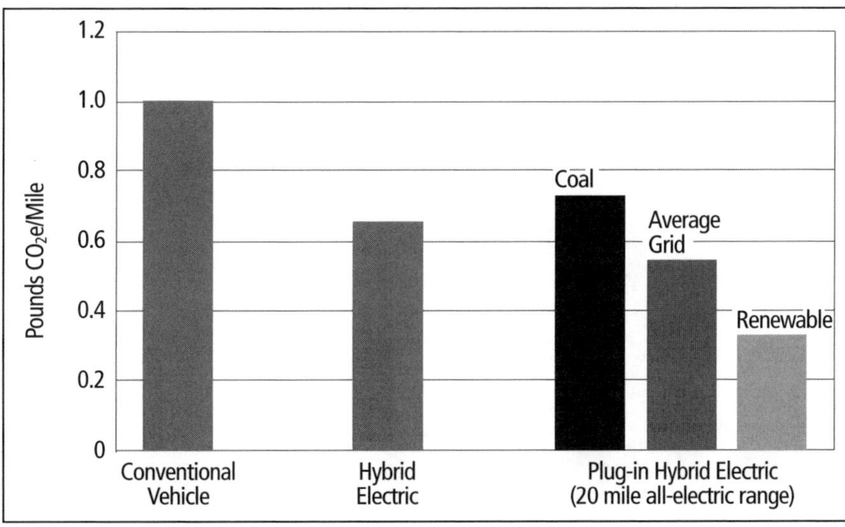

The PHEV is currently at a disadvantage compared to the ordinary hybrid when it comes to CO_2 reduction. The PHEV mpg and other electric car mpg claims have painted a far too optimistic picture. It should not be surprising to see only marginal change, considering that a car running on fossil fuels directly versus one running on fossil fuels via electricity is still a heavy object requiring energy for its movement. PHEV supporters focus on a contrived mpg rather than CO_2 generated obscured the marginal improvements likely.

Summary

Unfortunately, the PHEV became a brand before it became a product. The overstatement of its miles per gallon capability by the government, the mileage misrepresentation of a small number of ad hoc backyard Prius upgrades with no reference to the 34-mph maximum speed, together with a campaign now five years old saying the "technology is now ready," has probably delayed efforts for a more significant change in transportation modalities. It is important to realize that there are very few PHEVs in operation even at this time, in spite of the hype that began with Felix Kramer in 2002. A few hundred modifications of Prius cars[79] have been made – all operating under the 34-mph limit (finally fixed in mid-2009).[80] The PHEV is still a fossil fuel based car, dependent on fuels such as coal, natural gas, and uranium. It addresses the oil import problem rather than the climate change problem. Its utility is questionable and it may well follow the fuel cell automobile into obscurity.

Part II

The Next Phase

Introduction

The preceding section provided the historical context of electric vehicles and noted how long ago the technological concepts were developed. It emphasized the selling of the ideas and the misrepresentation of mileage claims (one measure of a vehicle's desirability). I recommend a more meaningful measure – the CO_2 generated.

This section analyzes the technology in some detail and points out some significant risks. The evolution of the EV1 to the Volt is discussed, noting that the Volt is not as revolutionary as portrayed. The actual performance of modified Prius cars is summarized, performance that is well below the expectations that have been created. The risks associated with a new battery technology are covered. Toyota's PHEV plan is analyzed in some detail as well as that company's concerns about the viability of pluggable hybrids in general.

The importance of scale is discussed which points out the difficulties of rapidly introducing new technologies. Energy systems evolve slowly as well as the products that use energy. The history of the Prius is covered, pointing out the relative small market penetration of this extremely popular and successful product. Several sources are sited that call into question the optimistic claims of delivery of millions of PHEVs in a relatively short time period.

Carbon Capture and Storage is important to the success of any form of electric car. The power grid must provide the electricity to operate the vehicles which means more coal and natural gas being burned. Presumably the CO_2 from power plants can be buried to reduce emissions but this is very problematic. Alternatively, renewables could be used to power the vehicles, but it is questionable if they can provide the volumes of electricity required. Another risky component of a successful transition away from gasoline is the ability of the power grids to provide the electricity to hundreds of millions of vehicles.

Understanding Hybrid Car Technology and Terminology

In a late-2008 news article, *Car Talk* hosts Ray and Tom Magliozzi wrote, "But if [we were] to simplify it, [we'd] say that today's gasoline electric hybrids use electricity to supplement a gasoline engine, whereas tomorrow's plug-in hybrids will use a gasoline engine to supplement electric power."[1]

This is a good layperson's summary but with a questionable assumption. It assumes that a pluggable hybrid electric vehicle (PHEV) is a follow-on, improved version of a contemporary hybrid, an idea fostered by advocates of the PHEV. In mid-2009 Toyota delivered the third-generation of the Prius, an improved hybrid car. Other manufacturers (including Toyota) will soon deliver versions of battery electric vehicles (BEVs), using lithium ion technology, repeating the BEV scenario of the late 1990s but with better electrical storage devices. In late 2009 Toyota delivered test versions of a PHEV with lithium ion batteries, using the third-generation Prius as the platform. Paraphrasing the Magliozzis' description slightly and removing the time progression (today and tomorrow), one might say that gasoline electric hybrids like the Prius use electricity to supplement a gasoline engine while electric vehicles like GM's Volt use a gasoline engine to supplement electric power.

This second wording suggests two competing technologies, one focused primarily on existing hybrid technology and the other on a new kind of car – a BEV with a gasoline motor assist. CalCars calls a PHEV a hybrid with a plug, as if something were overlooked earlier and now could be corrected. This simplified view provided PHEV advocates with "proof" that they were more innovative designers than the Toyota engineering department.

Prelude to the PHEV – EV1 with Auxiliary Power Unit

GM's first version of a PHEV was exemplified in an extension of the 1990s BEV EV1 to make it into a series hybrid unit.[2] (GM used the EV1 technology as a basis for additional electric car research.) It included lengthening the car to provide a back seat, as well as developing hybrid capability. An auxiliary power unit (APU) was added to the car, along with a small gasoline fuel tank. The APU was a gas turbine engine that started automatically when the battery charge dropped below 40% and delivered 40 kW of electrical power, enough to achieve speeds up to 80 mph and to recharge the car's NiMH (nickel-metal hydride) battery cells. Using the modified Magliozzi definition, this prototype was an "... electric vehicle... using a gasoline engine to supplement electric power."

Former GM president Rick Wagoner was quoted in an earlier report expressing his regret for axing the EV1 program and failing to put the right resources into hybrids. According to the March 13, 2007, issue of *Newsweek*,[3] "GM R&D chief Larry Burns... now wishes GM hadn't killed the plug-in hybrid EV1 prototype his engineers had on the road a decade ago: 'If we could turn back the hands of time,' says Burns, 'we could have had the Chevy Volt 10 years earlier.'"

GM has positioned the car as an improved electric vehicle stating,

> Unlike traditional electric cars, the Chevy Volt has a revolutionary propulsion system that takes you beyond the power of the battery. It will use a lithium-ion battery with a gasoline-powered, range-extending engine that drives a generator to provide electric power when you drive beyond the 40-mile battery range.[4]

This terminology could apply equally to the experimental EV1 with an APU. Basically, the Volt is using the same concept that Victor Wouk and Andy Frank experimented with decades ago. It is a BEV with an added engine – a competitor of the HEV rather than an improvement on it. Wagoner verified this when he noted in a 2008 *Economist* article, "And in 2010 GM is due to launch its revolutionary Chevrolet Volt, an electric car with a 'range extending' internal-combustion engine that promises to make the Toyota Prius look like yesterday's technology."[5] This argues for a succession but in actuality it is an alternative.

The development of the two technologies overlapped in time. The first EV1 (a BEV) was shipped in 1997 and the first Prius (a hybrid electric vehicle [HEV]) was shipped the same year. In the ensuing period the Prius has been phenomenally successful while the EV1 and other early BEVs have been canceled. GM hopes that adding a charging engine to a BEV will represent a breakthrough. Such a breakthrough may occur but it will be because of battery technology that is superior to the lead acid and NiMH battery technologies of the late 1990s, not because of a new design. But the same battery breakthrough will also be applied to the Prius, enabling it to operate partially as an electric vehicle. This may result in two mixed-mode cars: one a mostly gasoline-based vehicle with battery assist and possibly the ability to drive some distance on battery power (an enhanced Prius) and the other a mostly BEV vehicle with a gasoline assist (the Volt). It is unclear which technology will win out. But it is a mistake to think that the Volt represents a more advanced version of the Prius. The success of the Volt based on an untested configuration is not assured.

PHEV – Product or Concept?

The PHEV is more of an attractive concept than a product. In the last five years, a few hundred car modifications (90% of them to Priuses),[6] costing $7,000–$12,000 (one Ford Escape conversion cost $35,000),[7] have obtained some unknown and unmeasured electric driving range, operating at some unknown speed at some unknown reliability. A few hundred Priuses out of a population of over one million (the number of Priuses sold in the United States to date)[8] is hardly a huge endorsement by the population of hybrid owners. As mentioned in Part 1, the modified Prius was limited to a maximum of 34 mph in electric mode. The first conversion that could run at freeway speeds was not delivered until June, 2009,[9] years after CalCars had claimed the PHEV concept was proven. The idea of entrepreneurs in home garages modifying complex products like the Prius to develop an electric vehicle (EV) prototype is impressive, but without independent analyses and comparisons, it must be treated more as public relations than as real product development.

Larger organizations have monitored the progress of their plug-in cars. In May 2008, Seattle mayor Greg Nickels announced a plan to test plug-in hybrid electric vehicles in his city.[10] Seattle joined with the Port of Seattle, King County, and the Puget Sound Clean Air Agency to test the performance of PHEVs in an urban area. The Department of Energy's (DOE's) Idaho National Laboratory (INL) provided matching funds. Thirteen Priuses were converted at a cost of $12,000 each by A123Systems, a battery company located in Watertown, Massachusetts.

Nine months later (February 2009), in an article entitled "Reality check on plug-in cars" Danny Westneat discussed the results.[11] He noted that the cars were advertised as getting more than 100 or 150 miles per gallon but the actual results were 51 miles per gallon. (Recall the CalCars 100 mpg+ claim.) The plug-in Priuses were driven a total of 17,636 miles. They used standard gas/battery hybrid Priuses with an extra battery charged by using a 120-volt wall socket. Westneat also reported on Google's nine plug-in Prius hybrids that averaged 54.9 mpg.[12] He quotes Tom Turrentine, director of the Plug-In Hybrid Electric Vehicle Research

Center, in Davis, California as saying, "Google is now sort of embarrassed about their results. I think we all need to be more careful. When we say we're going to get 100 or 150 miles per gallon, then that's setting expectations way too high. It just leads to disappointment. We need to deal in reality." Westneat goes on to say that a lot of green projects will turn out to be hype.

There had been no independent analysis made of the successes and failures of the conversions of Priuses or any other car models for some years. In late 2009 an extensive study was conducted by independent researchers at the National Renewable Energy Laboratory (NREL) and INL.[13] This report discussed the performance of about 100 modified Priuses. The average mileage was reported to be about 51 mpg. Compare this number to the 150 mpg claim by President Obama and the 230 mpg claim by the then president of General Motors. This is an example of technical euphoria combined with enthusiastic people who make claims with no data or experience. The PHEV concept has been hawked for seven years by its promoters. The misrepresentation may have resulted in lost alternative opportunities. And a few hundred million more conventional gasoline cars have been built.

The Battery Question

New lithium ion batteries are a vital component of the PHEV. There is no unanimity regarding the performance, reliability, or costs of batteries for either new BEVs or PHEVs.[14] Menahem Anderman, a battery expert, says that a PHEV, due to the requirements of battery size, drive system, and vehicle design, is too expensive for the value. He notes,

> The environmental and societal benefit in moving from HEVs to PHEVs is smaller than that of moving from conventional vehicles to HEVs, but there is a much larger negative impact on consumer value. In moving from a PHEV to fuel cell HEVs, there is no additional benefit – and maybe even less. But the negative impact on consumer value is much higher. With battery electric vehicles, there is more benefit, but also more negative impact. The PHEV may be a long-term solution.

He concludes that PHEVs are unlikely to reach commercial volumes in the next seven years and that while it is not ready for commercial introduction now, the business risk in pursuing the platform now is tremendous. Anderman says that PHEVs are a detour and not a step forward,

> – if, for the sake of the PHEV, car and battery companies dilute their efforts

to expand conventional HEVs and to introduce lithium-ion batteries into the market;

– if governments miss the opportunity to provide incentives for conventional HEVs, "the only electrified vehicle technology that can make an impact on the environment in the next 10 years;" and

– if the PHEV is rushed to market, bypassing prudent automotive engineering design, verification, qualification, and supplier management standards.

Mark Duvall from the Electric Power Research Institute (EPRI), criticized Anderman's view, noting,

> What I'm saying is that we did a cost study, we think it's credible... and it disagrees dramatically with you. . . . There is greater leeway to discuss this. We can't assume the negative impact on consumer value. We have to look at the entire value equation of the vehicle.

According to Duvall the transportation sector has to do much more than develop plug-ins. Noting that HEVs would deliver only a very small portion of the total reduction required, Duvall declared that to say we should develop the HEV market fully and then pursue other avenues wasn't viable.

The Seattle and Silicon Valley (Google) results could bear out Anderman's warning. EPRI and other PHEV supporters may well be over-promising. If so, the unwarranted claims for the PHEV may haunt us in the future as we look at the years lost on this unproven idea. And the power industry will bear part of the brunt of public criticism. In late March 2009, when Wagoner was forced out of GM, the government announced that its analysis of the Volt was such that it did not appear to be competitive.[15] This may have been the prelude to cancellation of the project.

As mentioned in Part 1, battery cars have existed for 100 years. Hobbyists have been ripping apart conventional cars, replacing their engines with lead acid batteries for decades. The BEV efforts that were prompted by the California Air Resources Board (CARB) in the late 1990s and early 2000s began with lead acid batteries and part-way through the program changed to NiMH. There were key performance differences between the lead acid and NiMH, but high cost and other limitations did not give the NiMH versions sufficient improvements to create a viable BEV market.

The renewed interest in BEVs is based on the potential of lithium ion batteries. Lithium ion promises to increase the range beyond the NiMH batteries that were the core technology for the BEV experiments of the late 1990s as well as the current hybrid cars. The United States lags in battery development and manufacturing and Asian competitors are increasing market share. In late 2008 Panasonic purchased Sanyo. Panasonic makes the NiMH batteries for the Toyota hybrid line and is developing NiMH packs through its Panasonic EV Energy Company joint venture. The acquisition of Sanyo strengthens Panasonic's position.[16] Toyota has a joint venture with Panasonic and is using that company's lithium ion technology in a test program for Prius plug-in hybrids.[17] The United States is attempting to catch up with major investments from the DOE, including a $2.4 billion award for advanced batteries.[18]

BEVs are not necessarily the same thing as PHEVs, and there is confusion in this area, which affects battery choices. It is not obvious that a lithium ion driven PHEV will prevail over a lithium ion driven BEV. Possibly, the lithium battery will remove limitations of the original EVs. But the HEV will remain a competitor. More than two million HEVs have been shipped, compared to 5,600 BEVs in the 1990s and early 2000s (only a few hundred remain in use).

The availability of lithium is an important issue. Will we change from a dependency on oil from Saudi Arabia to a dependency on lithium from Bolivia? In a report "Lithium 2 – Under the Microscope,"[19] the writers quote Donald Garrerr, who says,

> Lithium is a comparatively rare element, although it is found in many rocks and some brines, but always in very low concentrations. There are a fairly large number of both lithium mineral and brine deposits but only comparatively few of them are of actual or potential commercial value. Many are very small, others are too low in grade.

The report goes on to say that lithium carbonate products will be sufficient for only a small fraction of the future global market requirements for PHEVs and EVs, estimating possibly for a maximum of 1.5 million Chevy Volts. (There are about 750 million cars worldwide.) Mining lithium could have environmental consequences, and the geopolitical relations between the United States and Latin America, never very good, could be exacerbated.

Toyota's PHEV Plan

Toyota, the undisputed leader in hybrid technology, is not ignoring the PHEV. In July 2007 an experimental Prius PHEV was certified for public road use by Japan's Ministry of Land, Infrastructure, and Transport. It used a NiMH battery pack and had an all-electric range of eight miles with a maximum speed of 62 mph.[20] (Recall that all the U.S. modified Priuses were limited to a maximum speed of 34 mph until the summer of 2009.) Two units were brought to the United States for assessment at the University of California (UC) Irvine's Advanced Power and Energy Program and UC Berkeley's Institute of Transportation Studies.[21] After certifying the car in Japan, Toyota formed a partnership with EDF (Électricité de France) to evaluate PHEVs in Europe.[22] Road trials began in France in autumn 2007. The vehicles were based on the Prius with an expanded NiMH battery pack to be recharged from the grid. EDF and Toyota developed a charging and invoicing system, installed in each of the test vehicles, that is compatible with a new generation of public charging stations. A year later EDF and Toyota extended these trials to the United Kingdom.[23]

Toyota has taken a measured approach to its PHEV research, saying that the concept is not yet proven. Compare this to the attitude of EPRI, GM, and CalCars who early on claimed that the concept was already proven. The Seattle and Google experience with third-party modified versions of the Prius disproved that point. Note that Toyota did not provide a PHEV prototype until the car could run at a decent speed (62 mph) while CalCars and others marketed cars with a 34 mph maximum speed limit when operating from batteries.

In early 2008 Toyota's president, Katsuaki Watanabe, said the company would introduce a test fleet of several hundred plug-in hybrids using lithium ion battery packs made by Panasonic.[24] Toyota has continued to research PHEVs but refuses to set a date for marketing them until tests are completed and the concept is verified. This is a very different approach from that of GM, which began selling the Volt car long before it was built. Test PHEVs will go to governmental, academic, and commercial fleets in Asia, Europe, and the United States for large-scale field testing. In early 2009 EDF and Toyota expanded the PHEV demonstration efforts in Strasbourg, France.[25] This was to include the delivery of about 100 units of next-generation PHEVs equipped with lithium ion batteries. In June 2009 Toyota announced it would begin leasing 200 plug-in versions of its third-generation Prius equipped with lithium ion battery packs in Japan starting at the end of 2009.[26] An additional 150 will be leased in Europe, with yet another 150 in the United States. It is typical to lease experimental cars such as was done with the electric vehicles in the 1990s. Toyota says that these cars will be used for real-world evaluation of performance and monitoring of driver attitudes and experiences.

In September 2009 Toyota gave more details of the PHEV Prius, noting that it will support an all-electric driving range of about 12 miles with a maximum speed of 62 mph.[27] The lithium ion pack can be fully recharged in one and a half hours from a 230-volt power supply. The PHEV Prius is based on the new third-generation Prius delivered in 2009. This is a far less ambitious specification than is being touted for the Chevy Volt.

Toyota's PHEV/EV Concerns

Although Toyota has continued the R&D and testing of the PHEV, the company is still not fully committed to the technology. This is a reason for concern. GM's success at coming up with new ideas and implementations has been poor – recall the EV, the PNGV research vehicles, and the FCEV. In an article entitled "Toyota Concerned About Market Viability of Plug-ins, Sees Clear Path to Commercialization of Fuel Cell Technology in 2015,"[28] the author quotes extensively from a speech given by Michael O'Brien, Toyota's U.S. corporate manager for advanced technology vehicle planning, at the CARB's ZEV Technology Symposium in Sacramento, California on September 22, 2009. O'Brien noted that Toyota, with 15 years of experience with advanced battery technology and the now-mainstream Prius, has key unanswered questions regarding market acceptance of plug-in hybrid and electric vehicles. O'Brien said it is difficult to force technology adoption by consumers. He said that the current state of market readiness for PHEVs and BEVs presents serious challenges: "Creating consumer demand for mandated advanced technology vehicles will require substantial government engagement at all levels" and "Prius best embodies Toyota's philosophy on advanced vehicle technology – for vehicles to become true solutions to our climate change and environmental challenges, they must be mass marketed solutions. It is not enough to build niche vehicles that only a few can afford to buy or fulfill only a small subset of customer needs."

O'Brien noted that Toyota has collected extensive data using hybrids, plug-in hybrids, and electric vehicles in real-world testing, which has convinced the company of three things:

1. NiMH batteries appear to be the best choice for conventional hybrid vehicles for some time to come.

2. The higher energy density of lithium batteries will be required for electric vehicles (EVs) and plug-in hybrids. Cost will remain an issue for years to come and will not be offset by large-scale mass production volumes. In addition, compared to conventional hybrid vehicles, batteries for EVs and PHEVs will be exposed to very severe usage conditions and wide swings in state of charge. Toyota believes that customers may need to consider replacement of the original battery during the vehicle's life, which adds to the true cost of the vehicle.
3. There is a need for a battery beyond lithium ion capable of higher energy density, lower cost and more stable lifetime performance.

O'Brien also noted that the Prius had been launched in 2000 but it took eight years for it to achieve the same level of purchase intention as other Toyota models. He said that it wasn't until the introduction of the Gen 2 Prius, with improvements in size, comfort, and fuel economy that Prius grew beyond niche volumes. O'Brien referenced a study by Synovate (a global provider of market research information and analysis) that found the average additional amount consumers would pay for 15–20 miles of electric range is $1,700 – far below the cost of batteries that deliver that kind of range. And pure BEVs, with even larger battery packs, increase the challenges. Such vehicles may be limited in range. O'Brien reviewed the difficulties Toyota had in marketing the RAV4 EV and the early Prius version and suggested that there is a limited natural market for plug-in vehicles. He noted that strong incentives will be needed to push demand to the higher levels and to address challenges regarding battery technology, cost, and suitability.

New Contenders – China

In China, the BYD Company Ltd., originally a battery company,[29] formed BYD Auto Company Ltd. as a subsidiary. The parent company is the world's second largest rechargeable battery producer. BYD has more than 130,000 employees worldwide and a market valuation of $7 billion. Warren Buffett made a significant investment in the company, enhancing its brand recognition. BYD began research and development of electric-vehicle batteries in 1996. In late 2002 the company formally established the Electric Vehicle Project Department, launching an R&D program on pure electric vehicles and dual-mode (DM) electric vehicles. BYD's long history and leading position as one of the world's top battery manufacturers gives it the necessary experience to build the power train technology for electric cars. BYD announced its F3DM plug-in hybrid at the January 2009 Detroit auto show. The car has a one-liter gasoline engine with an electric motor and can go 60 miles on the battery charge. It can run on batteries or a combination of the gasoline engine and batteries. The combined range is 360 miles. BYD also announced the e6 electric vehicle, a pure BEV. The range was announced to be 249 miles on a single charge. Charging time is one hour. This is not a small lightweight car like the Think, but a mid-sized vehicle.

The U.S. Strategy – GM and the Volt

GM's Volt offers three features: First, it is a BEV – the wheels are turned by electricity from batteries similar to those in EV1 and other cars of the 1990s and early 2000s, which were addressed in Part 1. Secondly, it uses lithium ion batteries. And third, it has added a range-extender gasoline motor that recharges the batteries while driving. These new technological features mean that the development risk is high. GM is developing multiple technologies in parallel. And that means schedules and price estimates are suspect. In an article entitled "Electro-Shock Therapy" in the July/August 2008 *Atlantic*, author Jonathan Ranch describes the risks being taken by GM with the Volt. This is a "bet your company" situation and GM is in desperate straits financially.[30] In December, GM announced that the GM Volt engine plant in Flint, Michigan would be put on hold.[31] The car was scheduled for 2010, which will probably move quickly to 2011. New technologies need hundreds of cars on the road for testing. So far there are none nor has a plan been announced.

When Could the PHEV Be Deployed?

Considering Scale

In "Moore's Curse and the Great Energy Delusion"[32] author Vaclav Smil questions the idea of rapid development and deployment of energy systems and, by implication, new energy-using technology. Moore's Law, named after Intel co-founder Gordon Moore, describes a long-standing trend in computer processing power whereby a computer's power doubles every year and a half. Smil assumes that this may have led people to think that since "the price paid for the same performance came down by 50% every 18 months, year after year," something similar can happen with energy systems. But the performance of energy-generating systems such as power plants and energy-using systems such as cars improve performance between 1 and 2% per year.[33] Smil recalls the popular wisdom of the early 1970s that by now all coal plants would be gone and all electricity would come from nuclear fission. Breeder reactors were to be common. Today, coal still produces half of our electricity and there is no commercial breeder reactor operating anywhere in the world. In the early 1980s supporters of renewables suggested that by now we would get 30–50% of our energy from solar, wind, and biofuels. The actual number is less than 1%. Smil says that all these forecasts failed miserably because promoters ignored the inherently slow pace of energy transitions and that it is delusional to think that the United States can install in ten years sufficient wind and solar generating capacity to replace the power plants that took 60 years to construct.

Smil offers some numbers about energy transition. It took oil about 50 years to capture 10% of the global primary energy market and 30 years to go from 10% to about 25% of the total. The corresponding timeframes for natural gas were approximately 50 years and 40 years. Nuclear fission reached 10% of global electricity generation 27 years after the commissioning of the first nuclear power plant in 1956, and its share is now roughly the same as that of hydropower.

Smil commented on Al Gore's July 2008 proposal to "re-power" America, which included the following: "Today I challenge our nation to commit to producing 100% of our electricity from renewable energy and truly clean carbon-free sources within ten years. This goal is achievable, affordable, and transformative."[34] Even if such an enormous capacity addition could be accomplished in a single decade, the financial cost would be enormous. It would mean writing off the entire fossil-fuel and nuclear-generation industry, an enterprise whose power plants alone have a replacement value of at least $1.5 trillion (assuming at least $1,700/installed kW), and spending at least $2.5 trillion to build the new capacity.

Smil then goes into a similar analysis of the transmission requirements to bring wind from the Great Plains to the east and west coasts, as well as photovoltaic (PV) solar from the southwest to the rest of the country.

Having evaluated the time for power technology changes, he next considers the machines that use power. The gasoline-fueled internal combustion engine was first deployed by Benz, Maybach, and Daimler during the mid-1880s. The massive spread of automobiles began only during the 1920s and in Europe and Japan only during the 1960s. This is a process amounting to spans of 30 to 40 years in the U.S. case and 70 to 80 years in the European case between the initial introduction and decisive market success (with more than half of all families having a car). The first diesel-powered car (Mercedes-Benz 260D) was made in 1936, but it was only during the 1990s that diesels began to claim more than 15% of the new car market in major EU countries, and only during this decade (the first decade of the 21st century) that they began to account for more than a third of all newly sold cars. Once again, roughly half a century had to elapse between the initial introduction and significant market penetration.

Smil says that the time needed for such significant market penetration is mostly the function of financing, developing, and perfecting massive and expensive infrastructures. He says,

> For example, the world oil industry annually handles more than 30 billion barrels, or four billion tons, of liquids and gases; it extracts the fuel in more than 100 countries and its facilities range from self-propelled geophysical exploration rigs to sprawling refineries, and include about 3,000 large tankers and more than 300,000 miles of pipelines. Even if an immediate alternative were available, writing off this colossal infrastructure that took more than a century to build would amount to discarding an investment worth well over $5 trillion – but it is quite obvious that its energy output could not be replicated by any alternative in a decade or two.

Because of the nature of the energy system, it is unlikely that very rapid energy transitions will be realized. The current push for rapid development of various kinds of improved power plants, such as integrated gasification combined cycle (IGCC) and ultra-critical pulverized coal plants, as well as carbon capture and storage (CCS) and many varieties of full or partial electric vehicles, will not be realized in this decade and are unlikely to achieve significant market penetration in the next. Thus, PHEV or other electric car sales may not even achieve the Prius's market success, providing less than 1% of the world car population after the first decade. Scaling is not understood by the general public and probably not by most politicians, so backup plans should be put in place.

The Prius Scaling Example

The Prius is an excellent example of the scaling premise. It was announced in 1997 and first year shipments in 1998 totaled 18,000 units. The rapid buildup of all Toyota hybrid sales is shown in Figure 10.[35] Worldwide cumulative sales of the Toyota Prius passed the one million mark in April 2008.[36] It took ten years for Toyota to sell a million copies of an extremely well-designed efficient car in a world market of over 750 million vehicles.

Figure 10: Annual Global Sales of Toyota Hybrids

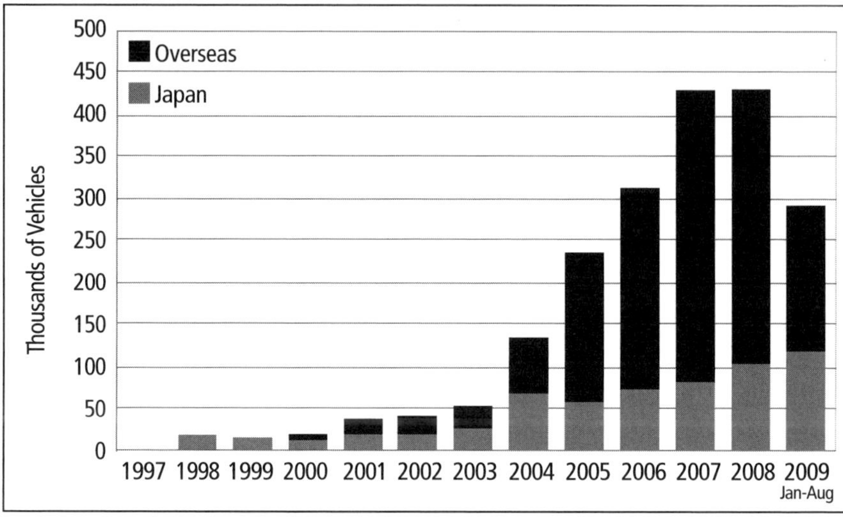

A realistic look at the potential sales of PHEVs shows a possible scaling scenario for that type of car. This was prepared by the U.S. DOE in its 2008 energy report.[37] The projection numbers for PHEVs are very small. (See Figure 11.)

In January 2009 the Boston Consulting Group published a report on potential sales of a variety of electric vehicles.[38] In this report they gave several perspectives on the success of the electricity-driven car. (See Table 5.) The "range extender" is another term for PHEV. Sales of the range extender are only 1.5 million vehicles in 2020.

In July 2009 Pike Research forecasted that a total of 1.7 million hybrids would be in use worldwide by 2015.[39] The United States is estimated to be the leading market with 610,000 vehicles sold over the next five years. China is estimated to produce about 540,000 during the same period. Toyota, GM, and Ford will account for the majority of sales.

This is a more positive view than the other references quoted. But it is important to realize that 350,000,000 conventional cars will be built during this five-year period, once more reminding us of scale and Smil's arguments relative to the speed of change.

Figure 11: Mild and Full Hybrid Systems Dominate New Light-Duty Vehicle Sales by 2030

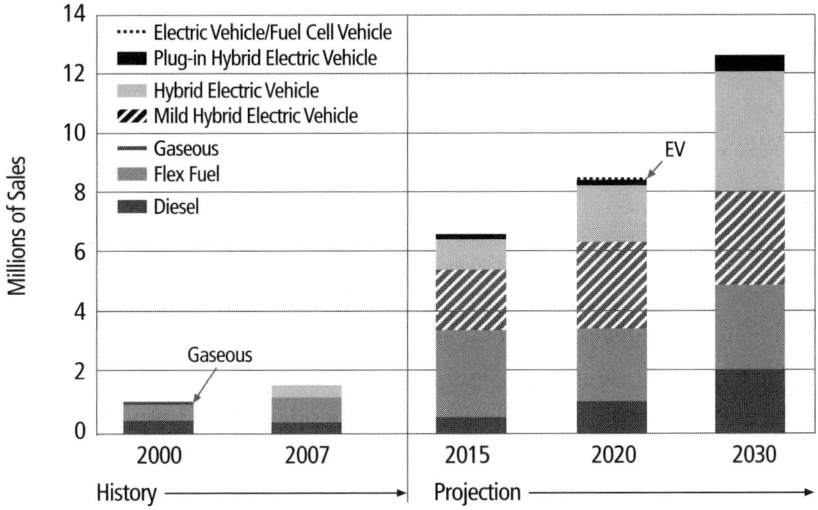

EIA Annual Energy Outlook 2009 Reference Case Presentation – December 17, 2008

Table 5: Car Sales in 2020 – Steady-Pace Scenario (Millions of Units)

	Western Europe	North America	Japan	China	Total
CNG	0.8	0.2	0	0.3	1.3
Electric	0.6	0.4	0.2	0.3	1.5
Range Extender	0.5	0.6	0.1	0.3	1.5
Hybrid	2.9	5.4	0.7	2.0	11.0
Diesel	5.5	1.2	0.2	0.9	7.8
Gasoline	6.7	12.9	3.4	8.4	31.4
Total	17.0	20.7	4.6	12.2	54.5

In July 2009 the DOE published its estimate of PHEV deliveries in the Smart Grid Systems Report.[40] This report was partially prepared by the Office of Electricity Delivery and Energy Reliability. Table 6 shows its estimate. This report and others illustrate the principle of scaling, which gives us a realistic timing. It shows 442,570 PHEVs in use versus the 1.7 million from the Pike report. Whatever the range, all such estimates show how long and difficult the path will be to convert from gasoline to electricity.

The media continue to support the idea of some kind of breakthrough. It is

Table 6: EV and PHEV Market Penetration

	EVs On Road		PHEVs On Road		EV Sales		PHEV Sales	
Year	Total in Use	% of Light-Duty Vehicles	Total in Use		Total Suites	% of Light-Duty Market	Total Sales	% of Light-Duty Vehicles
2006	28,891	0.01%		0.00%	173	0.00%		0.00%
2010	24,247	0.01%	35,526	0.02%	130	0.00%	35,526	0.26%
2015	17,840	0.01%	442,570	0.18%	149	0.00%	139,164	0.86%
2020	11,453	0.00%	1,322,438	0.51%	153	0.00%	237,212	1.43%
2025	6,787	0.00%	2,701,419	0.98%	165	0.00%	350,386	1.95%
2030	4,351	0.00%	4,282,767	1.44%	184	0.00%	443,207	2.21%

chic to refer to the Manhattan Project or the Apollo Program or the Marshall Plan, which cost $24 billion, $135 billion, and $550 billion, respectively, in current dollars. Another comparable undertaking is the interstate highway system, which ended up costing $425 billion in current dollars.[41] Replacing every car in the United States with a $40,000 Volt would cost about $9 trillion.

Carbon Capture and the PHEV/BEV

The PHEV success is not simply a matter of a new car development implementation, a new battery, and a new price. It is tied into different sources of energy – coal and natural gas – both of which generate CO_2. The electric vehicle, either PHEV or EV, is fundamentally a coal/natural gas car. Coal generates more CO_2 per BTU than any other option; oil is second and natural gas third. The long-term proposal for coal and natural gas generated electricity is to somehow capture the CO_2 that comes from the coal or natural gas power plant and bury it in the ground.

The research in this area started almost two decades ago. The first International Conference on Carbon Dioxide Removal was held in March 1992 in Amsterdam, and the second was held in Kyoto in October 1994. The third conference was held at the Massachusetts Institute of Technology (MIT) and hosted by the DOE and EPRI in 1996. The name was changed to International Conference on Greenhouse Gas Control Technologies (GHGT) and the fourth conference (called GHGT 4) was held in 1998 in Interlaken, Switzerland. GHGT 5–8 were held in alternate years and GHGT 9 was held in 2008. Nine conferences since 1992 imply a strong commitment to the concept.[42]

In July 1993 the final report of a document entitled "The Capture, Utilization and Disposal of Carbon Dioxide from Fossil Fuel Power Plants" was published by the staff of MIT's Energy Laboratory under DOE sponsorship. The document is a detailed and encompassing analysis of CCS (carbon capture and storage) and shows how much research has already gone into the technology. Considering the first conference in 1992 in Amsterdam and the 1993 release of this report, one can assume that the world has been studying the problem of CCS for about 17 years.

Carbon storage was first demonstrated in 1996 at Norway's StatoilHydro Sleipner site in the North Sea. Since 1996 Sleipner has pumped about one million metric tons of CO_2 a year into the seabed. A second project in the Snohvit gas field in the Barents Sea stores 700,000 metric tons per year. A third effort, the Weyburn project, is currently the world's largest carbon capture and storage project. Begun in 2000, Weyburn is located on an oil reservoir discovered in southeastern Saskatchewan, Canada in 1954. The CO_2 for this project is captured at the Great Plains Coal Gasification plant in Beulah, North Dakota, which has produced methane from coal for more than 30 years. The CO_2 is used for enhanced oil recovery with an injection rate of about 1.5 million metric tons per year. A fourth site is Salah, which, like Sleipner, is a natural gas reservoir; Salah is located in Algeria. The CO_2 will be separated from the natural gas and reinjected into the subsurface at a rate of about 1.2 million metric tons per year. Fourteen years after the first CCS installation, approximately 45 million metric tons have been pumped into the ground. During that time humans generated about 90,000 million metric tons.

The first U.S. effort at power plant CCS was called FutureGen. It was initiated in 2003 by the Bush administration, who terminated it in early 2008.[43] It was restarted by the Obama administration in 2009.[44] China may lead the world in CCS;[45] it gets 78% of its electrical energy from coal versus the United States' approximate 49%. Coal generates 30% more emissions per unit of energy than oil and 60% more emissions per unit of energy than natural gas. China creates 366g/kWh from coal while the world creates 320g/kWh. In terms of coal, China is less efficient by about 15%, which is a small amount relative to the rapid yearly growth of its coal-fired power plants. China's reputed "dirty" coal plants are only 15% less efficient than those in the United States. In the popular press, China is demonized for its dirty coal plants but actually U.S. plants are not much cleaner.

China began a GreenGen program that was intended to duplicate the U.S. FutureGen program. The objectives of the GreenGen program are to conduct advanced R&D on integrated coal gasification, hydrogen production, power generation, and CO_2 storage, very similar to the objectives of the corresponding U.S. FutureGen program. The goal is to achieve high coal-based generation efficiency and near zero emissions of all pollutants, including CO_2. It will verify key technologies and system integration technology. Finally, it will conduct a commercial demonstration at acceptable electricity prices and realize sustainable development of coal-based electricity generation. The technological goals are to complete an industrial demonstration project of 400 MW before 2020, with generation efficiency up to 60% and with over 80% of the CO_2 separated and treated.

The Chinese organization China HuaNeng Group (CHNG) approved GreenGen in 2004, a year after the U.S. FutureGen project was begun. The feasibility study of the first step demonstration project was finished in September 2006, construction began in 2007, and the initial operation of a 250-MW plant was scheduled for 2009. CHNG set up a GreenGen Industrial Alliance and invited the energy companies of the world to participate. A membership agreement was signed between CHNG and the FutureGen Industrial Alliance in 2005. The U.S.-based Peabody Energy Company, the world's largest private-sector coal company, joined the alliance in 2007. Plant completion was scheduled for 2009 although complete CCS is some years later.

Germany is also ahead of the United States.[46] In 2008 a small coal plant was built under contract by the Swedish power company Vattenfall. This is a 30-MW plant, quite small by normal power plant standards, that uses a technique called oxyfuel rather than the more common IGCC technology used in GreenGen and planned for FutureGen. The plant is operational but releasing CO_2 into the atmosphere. The company has not yet received permission to pump the CO_2 underground. Protests have occurred, attempting to stop the project, with citizens expressing their concerns about safety.[47] Naturally, utility companies argue that CO_2 is not dangerous. However, massive carbon dioxide gas releases from the African lakes of Monoun in 1984 and Nyos in 1986 killed nearly 1,800 people.[48] The safety of storing CO_2 in the gigantic amounts needed to limit its emissions has not been adequately evaluated.

Like many so-called breakthrough projects, CCS has a long history. Looking at early conferences held in 1992 and major papers in 1993, the estimate for the possibility of the first power plants coming on line in 2020 shows the long lead time of high-risk technologies. This also says little about deployment. By the time the first plant can be demonstrated, there may be dozens or hundreds more standard pulverized coal plants built, with a life span of 50 years or more. CCS may have no real effect on global warming CO_2 generation from power plants for another 20 years – and there is no guarantee that significant progress will be made in that period. Smil emphasizes the decade-long transitions of energy systems. So far, today's CCS efforts appear to be following similar time periods.

PHEV and Renewables

The nation spent several decades pursuing the fuel cell car. It never made sense in its early implementations, since the hydrogen came from natural gas. Proponents argued that this was only an interim step until renewables were available. This argument is being presented again in a slightly different form. In terms of the PHEV and EV, which are essentially coal/natural gas cars, it is argued that CCS will be needed at the beginning stage until wind and solar PV can be deployed sufficiently so that CCS will not be needed. But just as with the fuel cell car, one technology is dependent on its success for another to succeed. So the PHEV and EV must eventually run on wind and solar PV.

This returns us to the scale argument discussed above. Certainly EV cars have already been proven, as shown by the GM EV1 and other cars developed for California in the 1990s. Since Toyota is now developing a PHEV based on the Prius, one can assume that both PHEV and EVs will be deployable at some point – the technology is really not an issue. Of course, the battery question is not so clear, which itself is a scaling issue rather than a development issue. As stated earlier, there is no guarantee that a sufficient number of batteries can be provided for a deep penetration into the market.

Table 7 shows a seven-year history of energy sources for the United States. The average annual use has been around 100 quads (quadrillion BTUs). Because the total is about 100 quads the entries can be seen as an approximate percentage of total energy. Thus, solar PV is about 0.1% and wind about 0.5%. Combined, they are 0.6% in 2008, up from a combined 0.2% in 2004. But this is a very small number. And the idea of scaling this up by a factor of 20 or more in a short time period is very risky.

The America Wind Energy Association (AWEA) notes that 21,000 MW

Table 7: Energy Consumption by Energy Source, 2002 – 2008 (Quadrillion Btu)

Energy Source	2002	2003	2004	2005	2006	2007	2008
Total	97.9	98.2	100.3	100.5	99.9	101.6	99.3
Fossil Fuels	84.0	84.1	85.8	85.8	84.7	86.2	83.4
Coal	21.9	22.3	22.5	22.8	22.4	22.7	22.4
Coal Coke Net Imp.	0.1	0.1	0.1	0.0	0.1	0.0	0.0
Natural Gas[1]	23.6	22.9	22.9	22.6	22.2	23.6	23.8
Petroleum[2]	38.4	38.8	40.3	40.4	40.0	39.8	37.1
Electricity Net Imp.	0.1	0.0	0.0	0.1	0.1	0.1	0.1
Nuclear Elec. Power	8.1	8.0	8.2	8.2	8.2	8.5	8.5
Renewable Energy	5.9	6.2	6.3	6.4	6.9	6.8	7.3
Biomass[3]	2.7	2.8	3.0	3.1	3.4	3.6	3.9
Biofuels	0.3	0.4	0.5	0.6	0.8	1.0	1.4
Waste	0.4	0.4	0.4	0.4	0.4	0.4	0.4
Wood Derived Fuels	2.0	2.0	2.1	2.1	2.2	2.1	2.0
Geothermal Energy	0.3	0.3	0.3	0.3	0.3	0.3	0.4
Hydroelectric Conv.	2.7	2.8	2.7	2.7	2.9	2.4	2.5
Solar/PV Energy	0.1	0.1	0.1	0.1	0.1	0.1	0.1
Wind Energy	0.1	0.1	0.1	0.2	0.3	0.3	0.5
Solar/PV Energy %	0.1%	0.1%	0.1%	0.1%	0.1%	0.1%	0.1%
Wind Energy %	0.1%	0.1%	0.1%	0.2%	0.3%	0.3%	0.5%
Wind/Solar PV %	0.2%	0.2%	0.2%	0.3%	0.4%	0.4%	0.6%

Source: Energy Information Administration

of capacity has now been installed in the United States, which will provide 60 billion kWh of electricity in 2009.[49] But as previously noted, total electric generation is about 4.1 trillion kWh annually. The current wind and PV capacity provides about 1.5% of all electricity generated. Although wind and solar have grown rapidly in terms of percent increase each year, the growth is from a tiny base. As installations increase, the growth rates tend to decline. Thus, getting a high percent is not guaranteed.

Ted Trainer, an Australian researcher, is one of the few people who have attempted to analyze the variability and dispatchability of wind and solar.[50 51] He gives an example of an Australian wind farm whose capital investment cost was 13 times that of a comparable coal plant. The lifetime cost of the fuel is

4.2 times the capital investment of the power plant. The wind turbines are still two to three times more expensive. A similar example for a solar PV alternative gives a cost figure of 12–14 times the cost of the coal plant, including the cost of the coal it consumes. Trainer points out that wind turbines and solar PV should actually be built to supplement conventional power plants, not to serve as replacements or alternatives to building future conventional power plants. Their virtue is in reducing the amount of coal burned, not in avoiding building coal plants.

The Smart Grid

The Smart Grid is a proposal to redesign and reimplement the nation's massive electrical power system. Some have said that the greatest engineering achievement of the 20th century was not the car or computer but the electrical grid. The current view of the grid is that it is "dumb" and there is great hope that by making it "smart" our energy problems will somehow be alleviated. The argument is that the system is old and out of date and needs to be modernized. Critics say that the existing system is still electro-mechanical rather than digital. But the tremendous growth in electricity consumption in the last several decades could not have happened with an antiquated structure. The idea that the current grid cannot handle growth reliably is arguable.

The need for a modernized grid is not that the current one is antiquated but rather that it must now provide a huge increase in electricity to replace gasoline as the primary fuel for automobiles. It is not so much modernization and expansion as it is a change in the basic design. There is a great deal of marketing hype dealing with the grid as "one way" and arguing about the need for two-way communication. A correlation is often made to the Internet and sometimes to the phone system. But the process of electricity generation is to take power from a few thousand power plants, running a few tens of thousands of generators, and distributing it reliably to 300 million people. It is not at all like a phone system or the Internet, which is a way of sending information among a large number of people. A more appropriate analogy might be the water system, which is also a one-way system. One could consider having all houses collect roof water and pump it back up the water pipes to reservoirs – an idea that makes little sense.

The most recent plan for the smart grid was described in a speech in September 2009 by energy secretary Steven Chu.[52] Chu mentioned the need to avoid blackouts and improve efficiency but emphasized that the major change is the need to develop massive new transmission lines carrying renewable power across the country, being accessed by millions of plug-in vehicles buying cheap energy at night and selling it back for a profit during the day. First steps are smart meters (meters that do not need to be read monthly by utility staff) and distribution automation systems. Chu is visualizing a transmission and distribution system that will have to deal with variable power, which implies wind, PV, and other intermittent renewable sources. Chu notes that this requires massive energy storage, well beyond today's capabilities. He acknowledges that capacity beyond 20% intermittent renewables will require baseline generation sources that will be idling as backup for when the wind or sunlight fails. Finally, he argues for using the hundreds of millions of batteries in the hoped-for electric vehicle to balance the grid at peak demand loads. (This has been labeled vehicle-to-grid [V2G] systems). The thought of the overhead of having electricity flow to and from hundreds of millions of cars as temporary storage batteries is mind boggling. The new smart grid will also manage household use, which will interlink the appliances and manage the demand changes as required.

Table 8 summarizes many of the smart grid metrics and applications in four columns. The first column is the metric title and describes the application. The second column is labeled "Type" and has two categories – build and value. Build describes attributes that are built in support of a smart grid while value describes the value that may derive from achieving a smart grid. The third column measures Penetration/Maturity, abbreviated as Pnt/Mtr. Its options and their abbreviations (in parenthesis) are nascent (nasc), low (low), moderate (md), and mature (matr). The fourth column, Trend, includes these same options plus high (high), declining (decl), flat (flat), and improving (impr).[53]

The smart grid is essentially the application of information technology to the existing grid in hopes of optimizing energy use. A second application for the smart grid is to provide temporary electricity storage. The third conceptual application is to extend the electrical sources for the grid, currently a few thousand power plants, to hundreds of millions of electricity generators. For example, there are over 100 million buildings, all of which might have solar panels on the roof. This is a huge undertaking and may not be successful.

"Smart" home appliances are ones with an embedded computer chip that can be controlled (turned on and off) by the computers of the power company. These

Table 8: Smart Grid Metrics and Applications

	Metric Title	Type	Pnt/Mtr	Trend
	Area, Regional, and National Coordination Regime			
1	Dynamic Pricing: fraction of customers and total load served by RTP, CPP, and TOU tariffs.	build	low	md
2	**Real-time System Operations Data Sharing:** Total SCADA points shared and fraction of phasor measurement points shared.	build	md	md
3	**Distributed-Resource Interconnection Policy:** percentage of utilities with standard distributed-resource interconnection policies and commonality of such policies across utilities.	build	md	md
4	**Policy/Regulatory Progress:** weighted-average percentage of smart grid investment recovered through rates (respondents' input weighted based on total customer share).	build	low	md
	Distributed-Energy-Resource Technology			
5	**Load Participation Based on Grid Conditions:** fraction of load served by interruptible tariffs, direct load control, and consumer load control with incentives.	build	low	low
6	**Load Served by Microgrids:** the percentage total grid summer capacity.	build	nasc	low
7	**Grid-Connected Distributed Generation (renewable and non-renewable) and Storage:** percentage of distributed generation and storage.	build	low	high
8	**EVs and PHEVs:** percentage shares of on-road, light-duty vehicles comprising of EVs and PHEVs.	build	nasc	low
9	**Grid-Responsive Non-Generating Demand-Side Equipment:** total load served by smart, grid-responsive equipment.	build	nasc	low
	Delivery (T&D) Infrastructure			
10	**T&D System Reliability:** SAIDI, SAIFI, MAIFI.	value	matr	decl
11	**T&D Automation:** percentage of substations using automation.	build	md	high
12	**Advanced Meters:** percentage of total demand served by advanced metered (AMI) customers.	build	low	high
13	**Advanced System Measurement:** percentage of substations possessing advanced measurement technology.	build	low	md
14	**Capacity Factors:** yearly average and peak-generation capacity factor.	value	matr	flat

	Metric Title	Type	Pnt/Mtr	Trend
15	**Generation and T&D Efficiencies:** percentage of energy consumed to generate electricity that is not lost.	value	**matr**	**impr**
16	**Dynamic Line Ratings:** percentage miles of transmission circuits being operated under dynamic line ratings.	build	**nasc**	**low**
17	**Power Quality:** percentage of customer complaints related to power quality issues, excluding outages.	value	**matr**	**decl**
	Information Networks and Finance			
18	**Cyber Security:** percent of total generation capacity under companies in compliance with the NERC Critical Infrastructure Protection standards.	build	**nasc**	**nasc**
19	**Open Architecture/Standards:** Interoperability Maturity Level – the weighted average maturity level of interoperability realized among electricity system stakeholders.	build	**nasc**	**nasc**
20	**Venture Capital:** total annual venture-capital funding of smart-grid startups located in the U.S.	value	**nasc**	**high**

appliances also will be able to communicate with each other, that is, the toaster can "talk" to the washing machine. It's hard to determine the utility of this information management model. Whether energy will be saved is an open question. The idea of people watching their "dashboards" and bidding for lower prices for using the microwave is questionable. Utilities could measure energy constantly on a moment-to-moment basis and price on a moment-to-moment basis. Most of the smart grid applications are simply to suggest small gains in efficiency due to humans making buying decisions based on "smart" information. This will add one more level of complexity to our electronically burdened lives.

In October 2009 President Obama announced a $3.4 billion investment in smart grid technologies.[54] The investment was part of the American Reinvestment and Recovery Act. It will be matched by industry funding for a total public-private investment worth more than $8 billion. The announcement included a note that an analysis by EPRI estimates that the implementation of smart grid technologies could reduce electricity use by more than 4% by 2030. This is an astonishingly small amount when considering the depth of the cuts needed to avoid global warming. The major elements in the announcement and the investments included:

Empowering Consumers to Save Energy and Cut Utility Bills: $1 billion. This covers smart meters and other systems that allow consumers to access dynamic pricing information and to save money by programming smart appliances and equipment to run when rates are lowest.

Making Electricity Distribution and Transmission More Efficient: $400 million. Deploying digital monitoring devices and increasing grid automation to increase the efficiency, reliability, and security of the system, and support intermittent renewable energy resources.

Integrating and Crosscutting Across Different "Smart" Components of a Smart Grid: $2 billion. Projects that will incorporate various components into one system or cut across various project areas – including smart meters, smart thermostats and appliances, syncrophasors, automated substations, PHEVs, and renewable energy sources.

Building a Smart Grid Manufacturing Industry: $26 million. These investments will help expand the U.S. manufacturing base of companies that can produce the smart meters, smart appliances, smart transformers, and other components for smart grid systems in the United States and around the world.

The usefulness of all this is unknown. Any grid that can support today's massive electricity loads is modern enough. Automatic reading of meters is simply a labor-saving device. Consumers bidding for a few cents off their electrical bills is unlikely. The basic function is the support of the PHEV.

Summary

The PHEV is a combination of technology, hope, and hype. It has been justified by deliberately misleading ways of calculating mpg. The main perpetrator of the scam is the electricity industry through its industry organization EPRI. Very small advocacy organizations, either funded by EPRI or in agreement with its views, including CalCars, the University of California at Davis Institute of Transportation Studies laboratory, Plug In America, and Austin Energy, have sold Americans on a miracle technology. The government at all levels supports the electrification effort.

Honesty and full transparency calls for the mileage numbers of PHEVs to be given in terms of "miles per gallon equivalent" (mpge), similar to the way the energy industry uses barrels of oil equivalent (boe), to allow comparing all fuels, including coal, natural gas, and oil. Mpge is difficult to determine, except on a national basis, since fuel mix varies by region; nevertheless, it can be determined. An alternative to mpge is the use of CO_2 per mile. All cars should have a CO_2 per mile value, using the national fuel mix that would be applicable to every kind of drive train – gasoline, electric, or combined. This would eliminate much confusion and provide a true comparison.

The PHEV hype would have no impact except for the arrival of a potential new battery technology based on lithium ion chemistry. The evolution for vehicles from lead acid to nickel hydride and now to lithium ion represents very significant improvements in battery state of the art. This is rare in automotive development, which is usually incremental. If the lithium ion battery provides a *reliable* increase in capacity and decrease in charging time, then electric vehicles that did not succeed in the late 1990s, such as the EV1, RAV4 EV, and Think, may succeed now.

It is most likely that there will be two lines of vehicles in the future, the gasoline-based hybrid, such as the Prius, and the electricity-based vehicle, such as new offerings from China (BYD) and Japan (Nissan). The PHEV as an alternative third path may be much less successful because of the need to have two drive trains as well as multiple battery capacity points, e.g., PHEV 10, 20, and 40. No private car technology – gasoline, hybrid, diesel, PHEV, or EV – offers any promise for reducing the CO_2 emissions that may destroy the planet. Producing an increasing number of more efficient cars is a step in the wrong direction.

The risk is high. First, the cars must be built. Sufficient supplies of rare materials must be considered. The electrical grid must be extended into functions that may or may not be possible. Tens of millions of wind generators will be needed. Carbon Capture and Storage (CCS) must be proven to work. What is more likely is that we will build a small number of EVs and PHEVs, power them by coal, getting us off Middle East oil over which our control is limited. (We can control South American lithium supplies much more easily if need be.) This will improve our so-called energy security but our climate security will be compromised.

The Future of Cars

For the past century, the automobile has defined much of our culture. The private car connects us to work, food, and other goods, entertainment, education, love, and sex. It may have done more to destroy what we have termed community than any other invention. However, the life span of the car is limited. American automobile companies have gone in and out of bankruptcy; yet they still want to build large cars. The automobile industry has never addressed energy and CO_2 in a meaningful way. To provide support to build more destructive devices knowing the threat of climate change is unwise.

The collapse of the Detroit companies will not have that much effect on our mobility. Our inventory of over 200 million vehicles will continue to provide transportation. We may have to forego styling changes and impressing our neighbors; maintaining the existing vehicles a bit longer should be the norm. In Cuba, many American vehicles built in the 1950s are still on the road. That country transitioned to bicycles, buses, and cabs, reducing private car demand.

It is dangerous to think that we can replace our current fleet of cars and trucks with electric ones – full or partial PHEVs. There may not be sufficient resources; mass production of lithium batteries may soon exhaust the available materials, assuming these batteries ever reach production levels. Carbon caps or taxes may limit production. Consumers in bad financial straits, the case for most of us now, may not buy replacement vehicles. There may be a shift away from everyone over 16 owning a car to a limitation based on income level. Cars may become smaller. Jitneys and buses may increase in numbers. We may be clever enough to follow China and develop the electric bicycle. Or some other country (e.g., Ireland) may complete development of shared transport products that we can purchase. Lower speed limits may increase the useful life of existing cars and make electric biking less dangerous. The shift from lots of cars to van pools and buses may make transport safer on the whole. Government should heavily tax the private car at every stage – taxes on horsepower, breaks for jitneys and buses, constantly increasing taxes on gasoline and carbon taxes. A transportation system that uses a small fraction of the current energy should be one of the nation's top priorities.

Part III

The Smart Jitney: Rapid, Realistic Transport

Introduction

The world is facing the twin threats of declining fossil fuel resources and the global climate change caused at least in part by gases released from burning these resources. Automobiles play a major role in both of these potential catastrophes. The damage is already so bad that severe restrictions may have to be placed on consumption of the remaining fossil fuels, making alternative transport systems vital.

For many decades U.S. transportation has been focused on the energy-intensive private car. After World War II the United States made transportation via the private car its top priority – at the expense of public transportation. However, the private car, regardless of its convenience, can no longer serve as the principal mode of people transport. Its high cost, the depleting of fossil fuels, and climate deterioration, along with high rates of deaths and injuries, make it unacceptable and unsustainable.

Prudence requires a backup plan. A "jitney" is a small vehicle that carries passengers over a regular route on a flexible schedule. Basically, a jitney is a form of mass transit using cars and vans, not passenger buses. A "smart jitney" system could be developed rapidly and provide a very sizable (75%) reduction of both fossil fuels consumed and greenhouse gases generated by personal transportation in the United States. It could also be the model for a new and more efficient approach to personal mobility. At the very least, it could keep the U.S. economy functioning by giving people a way to get to and from work if suddenly there were insufficient fuel for private cars.

The Private Car Paradigm

The private car dominates our economy and our way of life. In spite of its numerous benefits, the many conveniences, and the sense of freedom associated with the automobile, we can't continue using a machine that has been so devastating for the planet. Traffic is worsening all over the world. As energy resources deplete and pollution worsens, even a 100–200 mpg automobile cannot be the main vehicle for billions of people.

Table 9: U.S. Transportation Fleet with Replacement Costs

Vehicle Type	Number	Median Life (years)	Cost to Replace 1/2 the Fleet (in 2003 $)
Automobiles	130 million	17	$1.3 trillion
Light trucks, SUVs, etc.	80 million	16	$1 trillion
Heavy trucks, buses	7 million	28	$1.5 trillion
Aircraft	8,500	22	$0.25 trillion

On average, every American buys 13 cars in his or her lifetime.[1] Table 9 shows the total U.S. transportation fleet except for trains.[2] Vehicles last a long time, so it would take many years to replace them all with more efficient ones, assuming highly efficient vehicles were available. Heavy trucks are extremely important for transporting food and other materials. In addition, the 8,500 aircraft flying at 30,000 feet do more climate damage than their numbers suggest because the emissions from burning fuel are deposited at higher altitudes. The size of the transportation fleet is huge and the investment is in the trillions of dollars.

Performance Improvements and Growth Trends

In his testimony to a Senate subcommittee, John German, Manager of Environmental and Energy Analyses for American Honda Motor Company noted,

> It is clear that technology has been used for vehicle attributes which consumers have demanded or value more highly than fuel economy.... If the current car fleet were still at 1981 performance, weight, and transmission levels, the passenger car CAFE [corporate average fuel economy] would be almost 38 mpg instead of the current level of 28.1 mpg. The trend is particularly pronounced since 1987. From 1987 to 2006, technology has gone into the fleet at a rate that could have improved fuel economy by almost 1.5% per year, if it had not gone to other attributes demanded by the marketplace.[3]

His testimony points out that the problems of energy use are not technical but cultural. Americans want speed, acceleration, and big cars – not efficiency.

Another reason petroleum consumption and accompanying CO_2 emissions are steadily increasing is the rapidly growing number of cars and trucks not just in the United States but worldwide. Figure 12 shows world growth projections for automobiles and petroleum fuels.[4] Note that the rate of growth

Figure 12: Projected Growth – Global Fleets and Petroleum Consumption

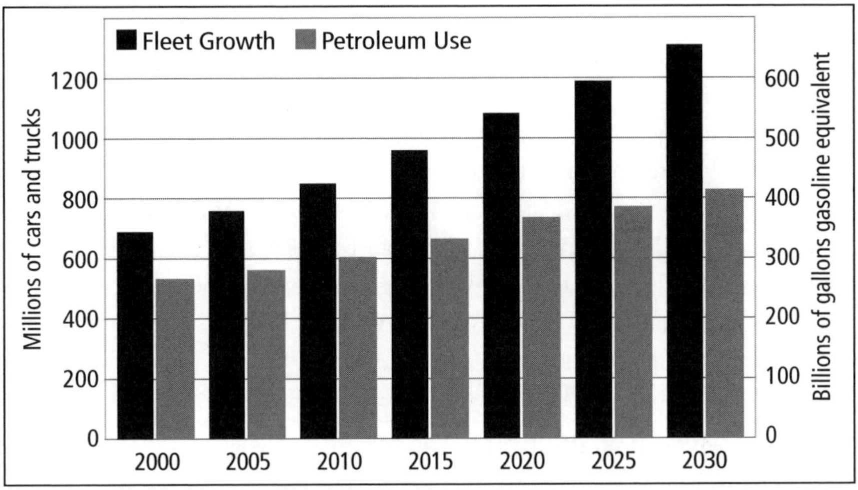

of petroleum (gasoline and diesel fuel) is slower than that of cars, reflecting some improvement in efficiency. But the total fuel consumption is increasing, making the case that greater efficiency does not reduce total consumption.

When oil production peaks, there will be an annual decrease in the availability of diesel fuel and gasoline. Those who argue that basic changes to the transportation system are not needed because of improved car efficiency are not addressing the implications of going from about 750 million cars today to nearly 1.3 billion cars in the next 23 years. Taking this a step further, today the average car in the world might get 35 miles per gallon with an average yearly mileage of 10,000 miles. In the future 1.3 billion cars getting 50–70 mpg and being driven an average yearly mileage of 15,000 miles will be disastrous to the planet. The increase in the number of cars and the distances traveled far outweighs the mileage improvements from any new type of engines. And, even if there were sufficient oil to fuel this growth, the amount of CO_2 generated would increase, worsening global climate change. Rapidly increasing the number of cars, while improving performance relatively slowly cannot continue. We must increase the passengers per vehicle as well as the miles per gallon.

Car Deaths and Injuries

The current car paradigm encourages people to take as many car trips as possible. Such heavy use of cars requires building and maintaining an enormous number of roads, garages, and parking areas. Advertising supports the cultural ideal of rugged individualism. People are encouraged to drive the largest possible cars while buses and trains become neglected alternatives. Walking and cycling can be inconvenient and dangerous because accommodating vehicles is given top priority.

The cost of the private automobile goes beyond financial and climate considerations. Figure 13 illustrates annual auto deaths and injuries.[5] The world total is about 1.2 million deaths and 40 million injuries yearly. In the United States about 40,000 people die each year in auto-related accidents. Deaths and injuries are also high in the developing world, where the infrastructure to support the car paradigm is not as well developed as in the United States. As cars begin to penetrate societies such as China and India, pollution, injuries, and deaths will increase there as well.

Figure 13: Worldwide Traffic Fatalities and Injuries – 1998

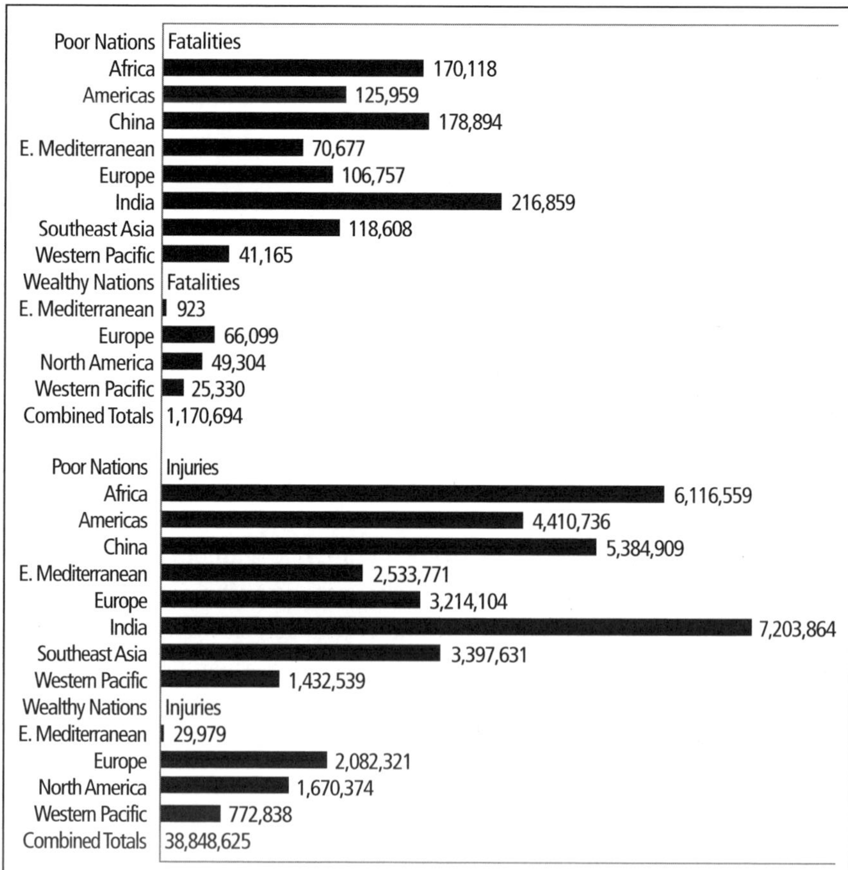

Results of the Private Auto Paradigm

Cars have played a major role in destroying community, by which I mean a feeling of home, family, and neighborhood. With the private automobile, we have gained speed and mobility but lost relationships with one another. We have chosen freedom for the individual over the integrity and support of the community. Of all our vaunted freedoms in the United States, none is more important than the freedom of the open road. Getting a license to drive is a rite of passage for our teenagers. The automobile allows young people to leave their communities, to experiment with high speed and, removed from family influence, with sex, alcohol, and drugs. We have accepted this so-called freedom for our children, ignoring potential damage and danger to their well-being. Parents' major fear is not teenage drug use or pregnancy, but death or injury in a car accident. And no wonder parents are concerned: almost every beginning driver in the United States has an accident. Parents don't want their children to be among the 40,000 yearly deaths or the 2 million injured, many with permanent disabilities.

Because people crave the freedom the car provides, they have accepted the destruction of communities, the negative impact on family life, and the deaths and injuries. But now the private car is threatening to destroy life on the planet as oil production begins to decline and climate change challenges the current transport paradigm.

The Limitations of Mass Transit

It is assumed that mass transit is a possible and obvious alternative to the private car. European cities are often praised for their superior transit systems. The New York City subway supposedly offers an alternative to the private car. But in all the cities with mass transit systems, the car population is still growing. Streets are becoming more crowded, and far more expenditures are made on roadways than on subways and buses. Subways and other forms of mass transit today only supplement the car since high density is required for mass transit. In successful past implementations, residential developments were laid out in dense corridors, typically along a rail or streetcar line, with open spaces and farms between these corridors. The ideal configuration was analogous to a wheel: the hub represented where people went to work and shop, while the spokes represented where they lived. The space between was often used to grow food.

When the private car became popular, the areas between the spokes were more accessible and were eventually filled in. Food growing was transferred farther from where people lived. This led to urban sprawl and suburbs, making effective mass transit more difficult. Eventually, there was no longer any attempt to build along mass transit lines, so many had disappeared. North American urban sprawl has no precedent in history, so the feasibility of a contemporary mass transit system has yet to be proven. A true mass transit system for the United States may, in fact, not be possible.

The potential energy savings of mass transit, in the context of implementing such a system in today's configuration of cities and urban sprawl, may be highly overrated. Table 10 depicts the BTUs of energy per passenger mile (assuming average passenger densities) for each type of transportation. This demonstrates that existing mass transit systems in the United States do not provide significant fuel savings. However, vanpool BTU per passenger mile is 1,322, hinting at the viability of a jitney system.[6]

Table 10: BTU Costs for Mass Transit vs. for Other Kinds of Vehicles

Private car	3,512
Light truck (SUV)	3,944
Vanpool	1,322
Bus transit	4,235
Airplane	3,261
Amtrak train	2,650
Rail commuter	2,996

The Smart Jitney Option

A new transport paradigm would place the highest priority on minimizing the use of fossil fuels, rather than on convenience, speed, or personal freedom. A smart jitney system would increase service and reduce energy use by more efficiently using existing vehicles. Jitneys typically are not required to travel exact routes on an inflexible schedule the way trains, buses, and street cars are. A U.S. jitney system could increase passenger occupancy from the current 1.6 persons per trip to three to five times that number. An increase of three times would use one third the number of vehicles to achieve the same number of passenger trips, removing most of the cars on the road and thus substantially reducing fuel consumption and CO_2 emissions.

The Smart Jitney Technology

The technology needed for smart jitney implementation is already available in the form of existing automobiles. Jitneys can be any vehicle, new or old, small or large, but with the minor addition of a special cell phone connected to the car. The cell phone would include GPS capability as well as an emergency call button for security. Whenever the rider or driver felt any sense of danger or threat, punching an emergency call button could automatically transmit information to the nearest law enforcement center for assistance.

Each passenger using the jitney system could use a personal cell phone, computer, or regular phone to access the system. Initially, there would be relatively small adjustments to the existing vehicle fleet and the cell phones currently dominating communication. Reservation tracking systems would need to be developed and installed. The reservation system would control both the ride management and bookkeeping of this new transportation modality. Rides would be planned and scheduled in a way similar to an airline reservation, except in a more timely, local, and responsive manner.

Scientists have already developed many kinds of ride optimization algorithms to coordinate complex pickup and deliveries for both people and materials. Systems used by FedEx and UPS routinely optimize pickup and delivery. An Auto Event Recorder (AER) is analogous to the flight recorder on an airplane. AERs already exist on more recently manufactured automobiles; the National Transportation Safety Board (NTSB) estimates that 65–90% of all vehicles in the United States contain some type of AER. These systems record driving activity, including vehicle speed, that is taking place in real time. This information would provide the basis for adding a new level of traffic safety and could be fully implemented nationally.

The Smart Jitney Process

The process begins with a request for service initiated by a passenger contacting the control center and entering a pickup location and a destination location along with desired times for pickup and drop off. One could also specify the level of service desired (see Options for Levels of Service below).

The smart jitney control center would constantly monitor information about all cars in the system, including the number of passengers, destinations, and vacant seats available. Once the analysis was completed (requiring only a few seconds of computer calculation), the rider would be assigned to a participating vehicle. The driver of the vehicle would be notified and provided the pickup location and time, along with directions.

The rider would be picked up and dropped off as requested. The driver would be compensated for providing the transportation service, with the fee regulated similar to mass transportation fees. After the trip, the rider would submit an evaluation of both the driver and the condition of the vehicle by cell phone or via the Internet, similar to the evaluation used by the Internet based company eBay. By publishing customer satisfaction for all to see, eBay eliminates many complaints because people simply stop buying from sellers with poor delivery performance. By having both a ride evaluation and AERs, driver records of long-term performance would be available.

Options for Levels of Service

The easiest and most efficient system is one in which all riders take whatever ride is available, but it might be difficult for Americans to share in such a completely democratic system. Therefore, different levels of service might be required. The first level of service could be more or less random. Only the pickup and destination locations would be entered along with the time of pickup and preferred time of drop off. The rider would input the data and the system would inform him or her of the car description, driver name, and time of pickup. This level would allow for the most possible rides and the quickest service.

A second level of service could allow a person to input preferences, requesting rides with certain groups of people. Possibly the most important would be for women to be able to request rides with other women. Men could also request non-coed trips. Another option would be to request certain age groups. Still others might want to put limits on the playing of radios or wish to ride with people who will be quiet. Any rider could be allowed to select the mode that best suits him or her. Of course, if a rider's preferences were too strict, availability of rides would decline.

A third level of service could allow scheduling future rides with a specific set of people. For example, a group of people with mutual interests who have a predictable schedule on a regular basis (such as work or school) could easily plan to travel together. Other levels of service would be added as experience dictates.

The Smart Jitney Driver

The smart jitney need not be implemented as a separate business like a taxicab service or a mass transit business. It is a form of ride-sharing using existing passenger vehicles and existing drivers. Overall, the number of people driving should decrease significantly. Although people could still drive and maintain an automobile, it is expected that eventually most people would accept the role of passenger. Anyone with a good driving record could serve as a jitney driver, but certain limitations would be stipulated. For example, inexperienced drivers could not be jitney drivers. Minimum age limits for drivers might correspond to age limits set by insurance and rental car companies, which reflect the higher accident rates of younger drivers.

More rigorous driving tests could be administered to grant qualification as a jitney driver. People with poor driving records, as measured by accidents and traffic citations, could also be barred from being smart jitney drivers. People with DWI convictions might not be permitted to be drivers until some time had elapsed or they had completed some type of re-qualification. People with child molestation records would be excluded. Similarly, existing smart jitney drivers could lose driving privileges because of poor driving or the use of intoxicants.

Smart Jitney Benefits

The smart jitney's first advantage is that it could be quickly implemented using the existing U.S. vehicle fleet. A jitney system would make it possible for people to continue to travel fairly long distances to work and school, and for other necessities, when gasoline shortages begin or when people realize that the deteriorating climate effects of CO_2 emissions can no longer be tolerated.

Secondly, a smart jitney system could solve some of the problems of the existing system. For example, it could be safer. Walking, cycling, and riding buses could be made more convenient and cars less convenient, reversing the trend of the last century. This differs from most of today's proposed solutions, which involve combinations of mass transit and using coal-generated electricity to fuel private electric automobiles.

A third benefit of the smart jitney is faster transit time. Commuters sitting on freeways in any large American city experience stop-and-go traffic at rush hour, averaging only a few miles per hour for much of the journey. A smart jitney system would eliminate most of the cars currently on the road, allowing much more rapid flow of traffic. Even time for stopping to pick up and drop off riders would be small relative to time spent in the current congestion!

The fourth major benefit is drastically reducing consumption of fossil fuels. Increasing the number of passengers per vehicle would provide a large reduction in fuel use. (An SUV getting 10 mpg but containing six passengers achieves the same mpg per passenger as a Honda Insight with just a driver.) This will help avoid economic disaster, lower the chances of worldwide wars over fossil fuel resources and, even more important, reduce CO_2 emissions substantially. This is a major step in eliminating the specter of global warming with its potential for massive disasters and loss of life.

The final benefit of the smart jitney is eliminating the tens of thousands of deaths and millions of injuries we currently accept as part of ordinary life. The proposed technology would include monitoring driving in real time and recording driver performance by AERs. Lives would be saved because of the associated decrease in traffic and because the best drivers would be at the wheel. In addition, lowering speed limits to save gasoline will lower the accident rate even more.

Addressing Concerns: Security, Safety, and Privacy

Evaluators of this smart jitney proposal typically are concerned with issues of personal security, safety, and privacy. In general, women are more concerned than men about security. Other issues deal with a feeling of loss, both of private time while driving and of self-esteem associated with ownership of a vehicle. Americans have been taught to believe vehicle ownership says something about who they are. The automobile has become far more than transportation – it represents the good life.

To be fair, the private automobile has also been responsible for the fears and concerns many of our smart jitney evaluators expressed. Cars certainly have made crime much easier: perpetrators can be miles away from the scene of the crime in minutes. Date rape becomes easier when two people are alone in a vehicle.

Moreover, the image of the private automobile, as presented in advertising, is typically one of power, speed, and force. Cars are sold on that basis, with strong emphasis on the individual rather than on the community. The poor record of young male drivers may be based more on driving with a certain machismo image in mind than from a lack of driving skill. The smart jitney could serve as a vehicle for cultural change as well as a new transportation modality. But to do so, security, safety, and privacy problems must be addressed.

Security

Security refers to the risk and danger from other people who, for whatever reason, may intend some kind of harm to our persons or psyches. Concerns about personal security are not trivial. The United States is a dangerous place, and its citizens are more violent than the majority of people in the rest of the world. Women have good reason for concern. "I wouldn't ride with a man," many female reviewers of the smart jitney have said, stating openly their fear of the violence in our culture. The ability to choose to ride only with other women must be part of the smart jitney system. American men, although feeling more secure than women, must also take the necessary precautions for living in a violent society. Children, too, must be protected from violence, bullying, or other anti-social behavior.

As noted earlier, at the completion of each ride, passengers could be asked to rate their smart jitney experience, covering such categories as the condition of the vehicle and the skill and suitability of the driver. With multiple passengers daily providing reviews, poor or unsuitable drivers could quickly be identified and their jitney licenses taken away. Obnoxious passengers could also be identified by the rating system.

Safety

Safety in relationship to automobiles refers to the accidents, deaths, and injuries that come from a myriad of causes, including auto and traffic equipment, roads, driver errors, and recklessness. Initially, to ensure vehicle safety, there would have to be mandatory inspection of vehicles for smart jitney licensing. Annual inspections would also be required.

At first, the smart jitney system would use existing cars, but eventually, as much higher-mpg vehicles became available, they would be replaced. Newer vehicles would be developed with a focus on safety rather than style. Instead of more car electronics for watching TV, accessing the Internet, or automatically parking the vehicle, collision avoidance electronics could be installed. Automobile companies have always given priority to speed, styling, and image over safety. However, if the forty billion dollars spent annually on automobile research and development were redirected from styling changes, then major safety improvements could be realized quickly. Furthermore, cars could be designed for longevity and ease of repair, which would contribute to reducing CO_2 emissions by minimizing the amount of embodied energy expended on the automobile fleet.

Traffic equipment and roads must be carefully evaluated, but they are not the main reason for accidents. Driver errors could be dramatically reduced by setting a lower speed limit; 45–55 miles per hour is the optimum speed for efficient performance of internal-combustion engines. Slower moving vehicles with higher passenger density would leave more of our streets available for bicycles and also reduce the risks involved in riding bikes in traffic. Through this shift to ride-sharing, the United States could set a high priority on reducing highway carnage. Legislation and market demand for safe jitney vehicles could force automobile manufacturers to improve safety standards.

Privacy

Protecting privacy means respecting people's need to maintain the confidentiality of their personal identification, including name, pictures, employer, or place of residence. A breach in privacy occurs when people intrude in our lives through inappropriate access to this personal data.

In modern times, real privacy is increasingly limited even though people have an illusion of it in their private cars and homes. Internet access and phone records, along with other private information, can now be purchased by businesses, corporations, and individuals. Government agencies such as the CIA, FBI, and NSA maintain civilian databases. Marketing of people's personal information is now acceptable and viewed simply as a business opportunity. Governments, particularly since 9/11, are collecting massive amounts of data on people.

With people sharing rides with strangers on a daily basis, protection of privacy will be a challenge. The smart jitney system could use the same methods of anonymity and protection as banks or any other institution promising confidentiality with the similar caveat that these institutions use. That is, they cannot absolutely guarantee that ride sharing information will not one day be inadvertently revealed or stolen. Infractions of privacy could be traced back in the ride-sharing system by reviewing ride records, and appropriate responses to violations might include denial of ride service.

Implementations

Implementing a system of this complexity would not be difficult. Although the smart jitney is different from existing ride and car-share systems, they serve as models for how technology could be developed. One rental car company, Zipcar.com, has an Internet system with some of the scheduling features and GPS tracking proposed for the smart jitney. In the United Kingdom, liftshare.com administers a ride-sharing program that matches riders and loads to cars and trucks around the country, using phone and Internet connections. Liftshare has more than 360,000 individuals and businesses as members. Mitfahrzentrale.de, based in Germany, offers ride-sharing throughout Europe to its 675,000 members.

The most significant implementation I have identified is the company Avego, headquartered in Kinsale, Ireland.[7] Founded by Sean O'Sullivan, a previously successful entrepreneur in mapping software, Avego released its first version of "shared transport" in late 2008. A second release will be available in the fall of 2010. The product was demonstrated to me in early 2009 at the Avego headquarters. The system uses an IPOD as the user communication device. The website for the product notes that "Avego Shared Transport enables private cars to become part of the public transport network by providing a marketplace for drivers to offer their unused seats to other people in real time." The system is well designed and extremely convenient both for drivers and passengers. The implementation was done with a relatively small programming team and its functionality could be duplicated easily, that is a few millions of dollars rather than the hundreds of millions of dollars spent on a single new car model.

A variety of transportation companies are beginning to use the Avego system. In February 2010, VPSI, the world's largest vanpool service provider, began to deploy Avego's system for its nationwide fleet of more than 5,000 vans.[8] Twenty-five million passengers use VPSI vanpools annually. This illustrates that the Avego system can be used in a variety of modes.

Shared transport has not yet been taken seriously at the U.S. Department of Transportation. That organization is still committed to the private car as the basis for transport. Mass transit supporters still argue for their approach to save energy and reduce traffic congestion but the country is now so spread out that it is well nigh impossible. It is expected that both car and bus positions will shift when shared transport products become more widely known and when electric vehicles fail to deploy rapidly.

In the meantime, early pioneers for smart jitney concepts remain active. Bob Behnke, who originally used the term, continues to advocate for the approach, convinced that Wi-Fi and other technologies now make it more feasible.[9] Park Woodworth continues to research ridesharing opportunities in Washington State and reports that Microsoft is suggesting ways to use Facebook-like features as a methodology for people to address the concerns of riding with strangers. He and Behnke co-wrote a paper describing a possible implementation.[10]

Long-Term Implications

The use of fossil fuels changed North America from a place of small local communities with limited mobility and resources to one of large urban and suburban concentrations with high mobility. Goods and food are shipped from thousands of miles away. This high mobility is based on the private car with its continuous use of significant quantities of gasoline. The advantages of this way of living, especially considering the effect on the climate, are becoming more and more questionable. Low-energy ways of living will need to include many changes, such as devolution from concentrated urban centers to smaller communities and local food supplies. Eventually, a pattern of smaller neighborhoods and towns with a focus on walking and bicycling will be more the norm. This is not a return to some previous period in human history. Most of the world is living this way now. Medical science and other advances will not be abandoned even if people drive less and share vehicles.

A primary cultural value of the developed world is competition; it is the key principle of our economic system. Cooperation, a key value in other cultures, is a sign of weakness in the United States. Walking or getting in a car with a stranger or riding a bus may be just an inconvenience in other cultures. For Americans, it is a threat to how we view ourselves. The depletion of fossil fuel sources and climate change will force us to be more cooperative. The smart jitney may start as a short-term emergency solution within the existing infrastructure, since it will allow us to keep similar patterns of living. Any inconvenience and discomfort experienced will be overshadowed by the possibility of stopping planetary degradation and its threat to basic survival. In the long run, the smart jitney could evolve into some mode of transportation not yet envisioned. It could help serve the larger physical community in the future as an intra-city mechanism for longer travel. It's possible that evolution of a jitney-based mass transit system, within the context of a decentralized local way of living, will naturally occur.

Conclusion

As I was completing this book in April, 2010, the EPA announced new higher mileage standards for automobiles, raising the U.S. fleet standard to 35.5 mpg by the year 2016. The agency did not define the performance or emissions for electric cars. A Wall Street Journal article pointed this out, referring to the earlier announcements by GM of a 230 mpg Volt and by Nissan of its 367 mpg Leaf.[11] The article noted a backlash occurred when GM and Nissan made these claims. The extent of backlash took the EPA by surprise. The article also refers to GM's "short lived viral marketing campaign." Part of the April EPA announcement exposed the unrealistic mpg claims being made for electric and PHEV vehicles.

This has been going on for forty years. Car companies, with U.S. government support, have offered a myriad of so-called "breakthrough technologies" – fuel cells, electric cars, diesel hybrids – which were to have arrived decades ago. Since these claims began in the 1970s, hundreds of millions of gasoline cars have been added to the world fleets. And car companies continue offering more technofixes, ignoring the failures of the last four decades.

Peak Oil and climate change call for a new approach, one that does not depend on the private car, which someday may be viewed as possibly the most destructive device ever made and the prime example of American consumer values. Sharing rides may be a traumatic change from using private cars. But the threat of energy shortages, loss of jobs, and life-threatening climate change provides motivation for trying out a new approach, one that may have long-term benefits for the environment. It is my belief that significant and positive personal gains in terms of time, safety, and economics will also be realized. Choosing this transportation approach might well save us from climate disaster and could prove to be superior to what we have now.

The technology (hardware and software) is developed, making it feasible to replace the private car quickly with smart jitney ride sharing, substituting low energy shared transport for private transport. Rather than waiting decades to improve car miles per gallon adequately to meet the climate crisis – an approach that has a high risk of failure – we can change our current individual car paradigm quickly with little risk. Such a shift will change (for the better) relationships between people and help restore community. And our public behavior may have to adjust (also for the better). Just as people cannot drive when drunk, they may not be able to ride when drunk. If all participants respected one another, a smart jitney ride could be a real pleasure as well as an energy-saving convenience. It could be an opportunity to learn new things and meet new people.

It's becoming apparent that the combination of all the technologies required to make the pluggable hybrid electric vehicle (PHEV) work have not advanced substantially in the last five years. The PHEV has been heavily marketed by special interest groups and the government – ignoring any concerns. The public is being overwhelmed with upbeat, positive – and misleading – propaganda. But the slow progress shows the components are risky. Creating a complete transportation system from these components is extremely risky.

Fissures are appearing in the structure. John Pederson wrote a telling article entitled *Plug-In Vehicles: The First Great Fraud of the New Millennium*[12] analyzing the options for battery allocation between PHEVs and EVs versus conventional hybrids (HEVs). His innovative contribution is to show that if the amount of lithium ion batteries for vehicles is limited, which is not unlikely, then they should be allocated to HEVs. He gives an example of one such allocation showing that the same amount of batteries will supply 1.5 million Nissan Leaf EVs, 2.25 million GM Volts or 24 million Prius'. (See Table 11)

Table 11: Different Scenarios for Limited Lithium Ion Availability

Car Model	Number of cars	Millions of gallons of gasoline saved	Millions of pounds of CO_2 saved
Nissan Leaf	1,500,000	600	9,403
Chevy Volt	2,250,000	765	12,859
Toyota Prius	24,000,000	3,840	97,344

Similarly, the success of CCS is not guaranteed. In a recent paper entitled *Sequestering CO_2 in a Closed Underground Volume*[13] Professors Christine Ehlig-Economides and Michael J. Economides challenge some of the basic concepts of carbon sequestration, arguing that it is based on an erroneous premise that the pressure feeding the carbon into a rock structure is constant. The authors note that eventually the rock would fracture and allow the carbon to migrate or escape. They conclude that CCS is not a practical means to provide any substantive reduction in CO_2 emissions. Their ideas, along with those of John Pederson, deserve consideration.

By the end of 2011 it may be clear that electric vehicles cannot reduce CO_2 emissions in any meaningful way for many reasons. It is vital that the embryonic option of shared transport be tested. Prototype ride sharing systems show significant savings of time and money and people are seeing how convenient they can be. Ride sharing offers a real possibility of making deep CO_2 reductions in time to save the planet.

Endnotes

Editor's Note: Unless otherwise indicated, all web addresses as accessed March 2010.

Part 1

1. Barack Obama Speech: New Energy for America; <http://www.barackobama.com/pdf/factsheet_energy_speech_080308.pdf>
2. "GM Hopes Volt Juices Its Future" by Sharon Terlep, *Wall Street Journal*, 081209; <http://online.wsj.com/article/SB124998537270122333.html>
3. The History of Electric Vehicles; <http://inventors.about.com/library/weekly/aacarselectrica.htm>
4. The History of Hydrogen; <http://www.hydrogenassociation.org/general/factSheet_history.pdf>
5. "Victor Wouk, 86, Dies; Built Early Hybrid Car" by Stuart Lavietes, *New York Times*, June 12, 2005; <http://www.nytimes.com/2005/06/12/national/12wouk.html>
6. "His pet PHEV: UC Davis' father of the plug-in hybrid electric vehicle says we're finally ready to drive green" by Cynthia Hubert, Sacramento Bee, Apr. 18, 2008. Accessed July 2009
7. *Plug In Hybrids: The Cars That Will Recharge America* by Sherry Boschert, New Society Publishers, 2006
8. GAO report on Electric and Hybrid vehicles funding, Sept. 1, 1977; <http://redbook.gao.gov/17/fl0080346.php>
9. "Heroes for the Planet – Geoffrey Ballard, In a Hurry to Prove the 'Pistonheads' Wrong" by Margot Hornblower, March 8, 1999; <http://www.time.com/time/reports/environment/heroes/heroesgallery/0,2967,ballard,00.html>
10. "Ford Joins in a Global Alliance to Develop Fuel-Cell Auto Engines" by Anthony Depalma, *New York Times*, December 16, 1997; <http://query.nytimes.com/gst/fullpage.html?res=9C00E3D7143FF935A25751C1A961958260>
11. "Daimler, Ford Buy Ballard's Auto Business" by Jennifer Kho, Nov 8, 2007; <http://www.greentechmedia.com/articles/daimler-ford-buy-ballards-auto-business-279.html>
12. *The Hype About Hydrogen: Fact And Fiction In The Race To Save The Climate* by Joseph Romm, Island Press, 2004
13. "GM, Toyota Doubtful on Fuel Cells' Mass Use" by Edward Taylor and Mike Spector, *Wall Street Journal*, Page B2, March 5, 2008; <http://www.energyinvestmentstrategies.com/2008/03/06/gm-toyota-reject-fuel-cells/>
14. "GM's $1.5 Million Hydrogen Fuel-Cell Vehicle: Won't Be Available for At Least 10 Years" by Adam Shake, 091508; <http://gas2.org/2008/09/15/general-motors-new-15-million-dollar-hydrogen-vehicle/>
15. *Powering the Future: The Ballard Fuel Cell and the Race to Change the World* by Tom Koppel, John Wiley and Sons, Canada, 1999
16. EV1 Electric: Series Hybrid; <http://www.autoworld.com/news/GMC/Series_Hybrid.htm>
17. EV1 Drivetrain Prototypes; <http://en.wikipedia.org/wiki/General_Motors_EV1#Drivetrain>
18. "Plug-in Vehicle History and Status: A Presentation to the California Air Resources Board, ZEV Technical Review Panel" by Plug In America, September 27, 2006; <http://www.pluginamerica.org/images/PIA_CARB_sep06.pdf>
19. Toyota RAV4 EV Description; <http://en.wikipedia.org/wiki/Toyota_RAV4_EV>

20. Toyota RAV4 EV Technology use in hybrids; <http://www.toyota.com/vehicles/rav4ev/>
21. Who Killed the Electric Car website; <http://www.sonyclassics.com/whokilledtheelectriccar/electric.html?detectflash=false&>
22. Partnership for a New Generation of Vehicles Description; <http://www.fueleconomy.gov/feg/pngv.shtml >
23. PNGV Discussion; <http://www.pngv.org/main/index.php?option=com_content&task=blogcategory&id=14&Itemid=26>
24. "A Detroit bailout must include a green makeover" by Jim Motavailli, *Guardian*, Nov 24, 2008; <http://www.guardian.co.uk/environment/2008/nov/24/network-travelandtransport>
25. GM Unveils Concept Car That Gets 108 Miles A Gallon by Jeff Kuhlman <http://www.electrifyingtimes.com/gmprecept.html>
26. Ford Prodigy by Jon Harmon, Jan. 20, 2000; <http://www.electrifyingtimes.com/fordprodigy.html>
27. Chrysler Hybrid-Electric Cars of the 1990 <http://www.allpar.com/model/intrepid-esx3.html>
28. Remarks by Vice President Al Gore on PNGV, The White House, March 30, 2000; <http://clinton4.nara.gov/WH/EOP/OVP/speeches/pngv_environment.html>
29. Review of the Research Program of the Partnership for a New Generation of Vehicles: Seventh Report (2001) by Transportation Research Board; *<http://www.nap.edu/openbook.php?record_id=10180&page=R16>*
30. CRS Report for Congress PNGV by Fred Sissine, Feb. 28, 1996; <http://ncseonline.org/nle/crsreports/energy/eng-9.cfm>
31. Freedom Car and Fuel Partnership Plan; <http://www1.eere.energy.gov/vehiclesandfuels/pdfs/program/fc_fuel_partnership_plan.Pdf>
32. Freedom Car and Fuel Partnership Plan, March, 2006; <http://www.hybridcars.com/history/history-of-hybrid-vehicles.html>
33. Toyota Prius Chronological History; <http://www.toyoland.com/prius/chronology.html>
34. "The Birth of the Prius" by Alex Taylor III, *Fortune*, February 24, 2006; <http://money.cnn.com/2006/02/17/news/companies/mostadmired_fortune_toyota/index.htm>
35. More Than 1 Million Prius Hybrid Cars Sold, May 20, 2008; <http://auto.moldova.org/news/toyota-sold-more-than-one-million-prius-hybrids-120558-eng.html>
36. "Honda Adjusts to Deteriorating Markets; Focus on Hybrids and Small Cars, Delayed Introduction of New Diesel in US and Japan" by Green Car Congress, 20 December 2008; <http://www.greencarcongress.com/2008/12/honda-adjusts-t.html>
37. "Honda president pushes founders energy, hybrid technology to ride out hard times" by Yuri Kageyama, Associated Press, July13, 2009; <http://www.sfgate.com/cgi-bin/article.cgi?f=/n/a/2009/07/13/financial/f044039D54.DTL>
38. Company Reports Automotive News, Accessed July 2009; <http://66.160.188.111/.eea05a1/cmd.233/enclosure..eea05a2> (Accessed January 2010)
39. "Toyota GM top global R and D spending, Industry reverses cutbacks of 2006" by Rick Kranz, *Automotive News*, Source: Booz & Co. Global Innovation 1,000 report, October 27, 2008; <http://findarticles.com/p/articles/mi_hb6674/is_200810/ai_n31226982/>
40. Fuel Economy Leaders: 2010 Model Year; http://www.epa.gov/fueleconomy/overall-high.htm

41. "On the Horizon: Building a Business Case, The Story in Brief" by John Douglas (Background info by Mark Duvall), EPRI History, *EPRI Journal*, Page 8, Spring 2008; <http://mydocs.epri.com/docs/public/000000000001016422.pdf>
42. EPRI's Contributions to the Development of PHEV Technology, November 2007 (Download Product ID Number 1014887); <http://et.epri.com/publicdocuments.html>
43. "Comparing the Benefits and Impacts of Hybrid Electric Vehicle Options," EPRI, Report 1000349 by multiple organizations, July 2001; <http://mydocs.epri.com/docs/public/000000000001000349.pdf>
44. "Driving the Solution-The Plug In Hybrid Vehicle" by Lucy Sanna, *EPRI Journal*, Fall 2005; <http://mydocs.epri.com/docs/CorporateDocuments/EPRI_Journal/2005-Fall/1012885_PHEV.pdf>
45. Calcars Site; <http://www.calcars.org/>
46. "What we can do and must do now to break America's Dependence on Foreign Oil and Bring Down Energy Prices" by Joseph Lieberman, October 7, 2005; <http://lieberman.senate.gov/index.cfm/news-events/speeches-op-eds/2005/10/e2809cwhat-we-can-and-must-do-now-to-break-americae28099s-dependence-on-foreign-oil-and-bring-down-energy-pricese2809d-loewy-lecture-georgetown-university>
47. "Oil and Security" by George P. Schultz and R. James Woolsey, A Committee on the Present Danger Policy Paper, August 5, 2005
48. *Plug In Hybrids: The Cars That Will Recharge America* by Sherry Boschert, New Society Publishers, 2006
49. "Gentlemen, Start Your Plug-ins" by R. James Woolsey, *Wall Street Journal*, 12/30/06
50. "Continuing the Clean Car Revolution" by Paula Dobriansky and R James Woolsey, *Wall Street Journal*, May 24, 2008
51. "Hybrid Vehicles Gain Traction" by Joseph J. Romm and Andrew A. Frank; *Scientific American*; April 2006.
52. "America needs to reduce its gasoline consumption. But how? By promoting plug-in hybrid electric vehicles" by Joseph Romm and Peter Fox-Penner, *Blueprint magazine* Electric Wheels, April 23, 2007; <http://www.ppionline.org/ppi_ci.cfm?knlgAreaID=116&subsecID=153&contentID=254272>
53. Austin Energy Company Description; <http://www.hoovers.com/austin-energy/--ID__54262--/free-co-profile.xhtml>
54. "Austin Alt Car: Austin Energy ends Plug-in Partners, unveils new smart charging software" by Sebastian Blancl, Oct. 18, 2008; <http://green.autoblog.com/2008/10/18/austin-alt-car-austin-energy-ends-plug-in-partners-unveils-new/>
55. Plug In America website; <http://www.pluginamerica.org/>
56. "Plug-in Vehicles are Practical, Proven, and Ready;" <http://www.pluginamerica.org/images/PIA_CARB_sep06.pdf>
57. Plug In America History; <http://www.pluginamerica.org/about-us.html>
58. LA Auto Show Loses Exhibitor in Fight Over Plug-In Hybrid Fuel Economy Promo; <http://blogs.edmunds.com/greencaradvisor/2008/11/la-auto-show-loses-exhibitor-in-fight-over-plug-in-hybrid-fuel-economy-promo.html>
59. "150 MPG Hybrid SUV Company Claims It is Being Muzzled" by Nick Chambers, Nov. 9, 2008; <http://gas2.org/2008/11/09/150-mpg-hybrid-suv-company-claims-it-is-being-muzzled/>

60. Chart showing AFS estimate of miles per gallon; <http://www.afstrinity.net/Jan08-AFSTrinity-XH150-Presentation.pdf>
61. Chevrolet Volt – GM's Concept Electric Vehicle - Could Nearly Eliminate Trips to the Gas Station; <http://www.batterypoweronline.com/images/PDFs_articles_whitepaper_appros/AppProCVolt.pdf>
62. RAV4 EV Mileage – Select 2002 and Electric; <http://www.fueleconomy.gov/feg/advancedSearch.htm>
63. Toyota RAV 4 History; <http://en.wikipedia.org/wiki/Toyota_RAV4_EV>
64. "EPA backs away from GM claim of 230 mpg for Volt" by Jeremy Korzeniewshi, Aug. 11, 2009; <http://green.autoblog.com/2009/08/11/epa-backs-away-from-gm-claim-of-230-mpg-for-volt/>
65. Internal-Combustion Engine Efficiency; < http://ffden-2.phys.uaf.edu/102spring2002_Web_projects/Z.Yates/Zach%27s%20Web%20Project%20Folder/EICE%20-%20Main.htm>
66. Smart Grid Systems Report, US DOE, Page 40, July 2009; <http://www.oe.energy.gov/DocumentsandMedia/SGSRMain_090707_lowres.pdf>
67. Energy Technology Perspectives, Scenarios and Strategies to 2050, International Energy Agency, 2006
68. Electric Power Statistics – See figure 2; <http://www.eia.doe.gov/cneaf/electricity/epm/epm_sum.html>
69. Electric Power Statistics – See Sales; <http://www.eia.doe.gov/cneaf/electricity/epa/epa_sum.html>
70. "Edison's Revenge," *New Scientist*, October 11, 2008; <http://www.transnational-renewables.org/Gregor_Czisch/projekte/NewScientist20081011.html>
71. National Labs Developing Methodology for Estimating Real World Fuel and Electricity Consumption of Plug-in Hybrids Based on Dynamometer Data by Green Car Congress, 29 September 2009; <http://www.greencarcongress.com/2009/09/nrel-phev-20090929.html#more>
72. EPRI History, *EPRI Journal*, Page 8, Spring 2008; <http://mydocs.epri.com/docs/public/ 000000000001016422.pdf>
73. IEA annual energy report; <http://www.eia.doe.gov/iea/carbon.html>
74. "Comparing the Benefits and Impacts of Hybrid Electric Vehicle Options," EPRI, Report 1000349, July 2001; <http://mydocs.epri.com/docs/public/000000000001000349.pdf>
75. Environmental Assessment of Plug-In Hybrid Electric Vehicles Volume 1: Nationwide Greenhouse Gas Emissions 1015325, EPRI Final Report, Page 7, July 2007; <http://mydocs.epri.com/docs/public/000000000001015325.pdf>
76. "Plug-in and regular hybrids: A national and regional comparison of costs and CO_2 emissions" by Eric Williams CCPP 08-04; Nicholas School of the Environment at Duke University; Nicholas Institute for Environmental Policy Solutions Center on Global Change, November 2008; <http://nicholas.duke.edu/ccpp/ccpp_pdfs/plug_in_hybrid.pdf>
77. "Plug-In Hybrids: An Environmental and Economic Performance Outlook" by James Kliesch and Therese Langer, ACEEE, Report Number T061, September 2006; <http://www.aceee.org/pubs/t061.htm>
78. "The Next Generation of Hybrid Cars: Plug-in Hybrids Can Help Reduce Global

Warming and Slash Oil Dependency," July 2007 NRDC; <http://www.nrdc.org/energy/plugin.pdf>

79. Locations of modified hybrids; <http://www.calcars.org/where-phevs-are.html>
80. "Software Upgrade to Prius PHEV NiMH Conversion Kit Enables All-Electric Driving At Up to 70mph" by Green Car Congress, June 10, 2009; <http://www.greencarcongress.com/2009/06/picc-20090610.html>

Part 2

1. "Plug In Hybrids Explained:" by Ray and Tom Magliozzi, *Dayton Daily News*, December 20, 2008; <http://www.spokesmanreview.com/auto/auto.asp?sid=438>
2. General Motors EV1 Discussion; < http://www.absoluteastronomy.com/topics/General_Motors_EV1>
3. Comin' Through *Newsweek*, March 13, 2007; <http://www.newsweek.com/id/36484/page/3>
4. Chevrolet Volt Site; <http://www.chevrolet.com/pages/open/default/fuel/electric.do>
5. "Mr Detroit," article about Rich Wagoner and Volt, *Economist Magazine*, Dec. 11 2008; <http://www.economist.com/people/displaystory.cfm?story_id=12758963> Reference July 2009
6. Locations of modified Prius cars; <http://www.calcars.org/where-phevs-are.html>
7. "Plug-In Hybrids Aren't Coming – They're Here" by Stuart Schwartzapfel, *Wired*, Oct. 5, 2008; <http://www.wired.com/autopia/2008/10/plug-ins-arent/>
8. "Toyota Global Cumulative Hybrid Sales Top 2M" by Green Car Congress, 4 September 2009; <http://www.greencarcongress.com/2009/09/tmc-2mil-20090904.html>
9. "Software Upgrade to Prius PHEV NiMH Conversion Kit Enables All-Electric Driving At Up to 70mph" by Green Car Congress, June 10, 2009; <http://www.greencarcongress.com/2009/06/picc-20090610.html>
10. Mayor unveils city of Seattle's first 100 mpg plug-in hybrid car; <http://www.seattle.gov/news/detail.asp?ID=8496&dept=40>
11. "Reality check on plug-in cars" by Danny Westneat, *The Seattle Times*, February 22, 2009; <http://seattletimes.nwsource.com/html/dannywestneat/2008771363_danny22.html>
12. Google site for their modified PHEVs <www.google.org/recharge/dashboard>
13. Deriving In-Use PHEV Fuel Economy Predictions from Standardized Test Cycle Results by J. Gonder and A. Brooker – National Renewable Energy Laboratory and R. Carlson and J. Smart – Idaho National Laboratory, Presented at the 5th IEEE Vehicle Power and Propulsion Conference, Dearborn, Michigan, September 7-11, 2009; <http://www.nrel.gov/docs/fy09osti/46251.pdf>
14. "To PHEV or Not To PHEV (At Least in the Near-Term)" by Green Car Congress, 17 February 2008; <http://www.greencarcongress.com/2008/02/to-phev-or-not.html>
15. "Government: Volt can't save GM" by Scott Burgess, *The Detroit News*, March 30, 2009; Referenced July 2009
16. "Panasonic Buying Sanyo for $9B; HEV and EV Batteries an Important Focus" by Green Car Congress, 19 December 2008; <http://www.greencarcongress.com/2008/12/panasonic-buyin.html>
17. Toyota FT-EV Concept 2009 Revealed with video, Prius plug-in hybrid production will

start later this year, by Michael Gauthier, January 11, 2009; <http://www.worldcarfans.com/9090111.009/toyota-ft-ev-concept-2009-revealed-with-video>

18. "DOE to Award Up to $2.4B for Advanced Batteries, Electric Drive Components, and Electric Drive Vehicle Demonstration/Deployment Projects" by Green Car Congress, 19 March 2009; <http://www.greencarcongress.com/2009/03/doe-to-award-up-to-24b-for-advanced-batteries-electric-drive-components-and-electric-drive-vehicle-d.html>
19. "The Trouble with Lithium 2 – Under the Microscope" by Meridian International Research, Les Legers, 27210 Martainville, France, May 29, 2008; <http://www.meridian-int-res.com/Projects/EVRsrch.htm>
20. "Toyota Announces Development of Plug-In Hybrid; First Manufacturer to Have PHEV Certified for Public Road Use in Japan" by Green Car Congress, July 25, 2007; <http://www.greencarcongress.com/2007/07/toyota-announce.html>
21. Toyota Delivers Plug-in Prius to UC Irvine, UC Berkeley, November 9, 2007; <http://pressroom.toyota.com/pr/tms/toyota/TYT2007110961930.aspx>
22. "EDF and Toyota Announce European Technology Partnership for Plug-in Hybrid Vehicles" by Green Car Congress Sept. 5 2007; <http://www.greencarcongress.com/2007/09/edf-and-toyota-.html>
23. "EDF Energy and Toyota Launch UK Trials of PHEV" by Green Car Congress, 10 September 2008; <http://www.greencarcongress.com/2008/09/edf-energy-and.html>
24. "Detroit 2008: Toyota President Katsuaki Watanabe pops in at party with surprise announcements" by Sam Abuelsamid, Jan. 13 2008; <http://green.autoblog.com/2008/01/13/detroit-2008-toyota-president-katsuaki-watanabe-pops-in-at-part/>
25. "EDF and Toyota Expand Plug-in Hybrid Vehicle Demonstration Efforts in Strasbourg, France" by Green Car Congress, 18 March 2009; <http://www.greencarcongress.com/2009/03/edf-and-toyota-expand-plugin-hybrid-vehicle-demonstration-efforts-in-strasbourg-france.html>
26. "Toyota to Lease 200PHEVs in Japan Starting at End of 2009, 500 Globally; Gen3 Prius Plus Li-ion Pack" by Green Car Congress, 3 June 2009; <http://www.greencarcongress.com/2009/06/tmc-200-phev-20090603.html>
27. "Prius Plug-in Concept to Support All-Electric Driving Up To About 12 Miles at Speeds Up To 62MPH" by Green Car Congress, 10 September 2009; <http://www.greencarcongress.com/2009/09/prius-phev-20090910.html>
28. "Toyota Concerned About Market Viability of Plug-ins, Sees Clear Path to Commercialization of Fuel Cell Technology in 2015" by Green Car Congress, 23 September 2009; <http://www.greencarcongress.com/2009/09/obrien-20090923.html>
29. "Detroit 2009: BYD's plug-in hybrids, the F3DM and F6DM" by Sebastian Blanc, Jan. 13, 2009; <http://www.autobloggreen.com/2009/01/13/detroit-2009-byds-plug-in-hybrids-the-f3dm-and-f6dm/>
30. "Electro-Shock Therapy" by Jonathan Rauch, *The Atlantic*, July/August 2008; <http://www.theatlantic.com/magazine/archive/2008/07/electroshock-therapy/6871>
31. "GM: Volt engine plant in Flint on hold; Flint Mayor Don Williamson says engine could be built at Flint Engine South," *The Flint Journal*, December 17, 2008; <http://www.mlive.com/flintjournal/index.ssf/2008/12/gm_volt_engine_plant_in_flint.html>
32. "Moore's Curse and the Great Energy Delusion" by Vaclav Smil, *The American*, November 19, 2008; <http://www.american.com/archive/2008/november-december-magazine/moore2019s-curse-and-the-great-energy-delusion>

33. *Plan C: Community Survival Strategies for Peak Oil and Climate Change*, by Pat Murphy, New Society, 2008
34. Al Gore Blog; <http://blog.algore.com/2008/07/a_generational_challenge_to_re.html>
35. "Toyota Global Cumulative Hybrid Sales Top 2m" by Green Car Congress, Sept. 4 2009; <http://www.greencarcongress.com/2009/09/tmc-2mil-20090904.html>
36. Worldwide Prius Sales Top 1 Million Mark, May 15, 2008; <http://www.toyota.com/about/news/product/2008/05/15-1-prius.html>
37. Annual Energy Outlook 2009, Early Release, Energy Information Administration, December 17, 2008; <http://www.eia.doe.gov/oiaf/aeo/pdf/aeo2009_presentation.pdf>
38. "The Comeback of the Electric Car: How Real, How Soon, and What Must Happen Next" by The Boston Consulting Group, authors Michael Book, Marcus Groll, Xavier Mosquet, Dimitrious Rizoulis, Georg Stichen, Jan. 2009; <http://www.bcg.com/impact_expertise/publications/files/Comeback_Electric_Car_Jan_2009_rev.pdf>
39. "Market Forecast Projects Cumulative 1.7M PHEVs Globally by 2015, 36% in the US" by Green Car Congress, 8 July 2009; <http://www.greencarcongress.com/2009/07/pike-20090708.html>
40. Smart Grid Systems Report, US DOE, July 2009, Page 27; <http://www.oe.energy.gov/DocumentsandMedia/SGSRMain_090707_lowres.pdf>
41. Interstate Highway System; <http://en.wikipedia.org/wiki/Interstate_Highway_System#cite_note-13>
42. The Third International Conference on Carbon Dioxide Removal, 30 April 1997; <http://web.mit.edu/energylab/iccdr3/info.html>
43. "Up in Smoke: President Bush's Big Environmental Broken Promise" by the Editorial Board, *New York Times*, Feb 13, 2008; <http://theboard.blogs.nytimes.com/2008/02/13/up-in-smoke-president-bushs-big-environmental-broken-promise/?pagemode=print>
44. "DOE revives FutureGen, Reversing Bush-Era Decision" by Ben German and Greenwire, June 12, 2009; <http://www.nytimes.com/gwire/2009/06/12/12greenwire-doe-revives-futuregen-reversing-bush-era-decis-47303.html>
45. Presentation – Near Zero Emission Coal Based Power Generation Project in China by Xu Shisen, Thermal Power Research Institute, Technology Development Center of CHNG, May 23, 2007; <http://belfercenter.ksg.harvard.edu/files/07-shisen%20xu.pdf>
46. "Clean-Coal Debut in Germany: New Coal Plant Is First to Capture and Store Carbon Dioxide" by Rob Edwards, Sept. 22, 2008; <http://abcnews.go.com/Technology/Story?id=5844357&page=1>
47. "Not under our backyard, say Germans, in blow to CO2 plan" by Terry Slavin and Alok Jha, *Guardian*, July 29, 2009; <http://www.guardian.co.uk/environment/2009/jul/29/germany-carbon-capture>
48. Volcanic Lakes and Gas Releases; <http://vulcan.wr.usgs.gov/Glossary/Lakes/description_volcanic_lakes_gas_release.html>
49. American Wind Energy Association notes top wind industry accomplishments of 2008. December 22, 2008; <http://www.awea.org/newsroom/releases/Year_End_Wrap_Up_22Dec08.html>
50. *Renewable Energy; What are the Limits?* by Ted Trainer, April 2004; <http://ssis.arts.unsw.edu.au/tsw/D74.RENEWABLE-ENERGY.html>
51. "Renewable Energy Cannot Sustain a Consumer Society" by Ted Trainer, Springer, 2007
52. "Chu Lays Out DOE's Smart Grid Vision, Standards to Come" by Jeff St. John, *GridWeek*,

September 21, 2009; <http://www.greentechmedia.com/articles/read/gridweek-chu-lays-out-does-smart-grid-vision-standards-to-come/>

53. Smart Grid Systems Report DOE, Page 8, July 2009; <www.oe.energy.gov/SGSRMain_090707_lowres.pdf>
54. "Obama Announces $3.4B Investment in Smart Grid Technologies and Projects" by Green Car Congress, 27 October 2009; <http://www.greencarcongress.com/2009/10/obama-smaratgrid-20091027.html>

Part 3

1. Toyota North America President Jim Press comments to 19th annual Automotive Outlook Conference, Detroit, January 9, 2007; <http://docs.google.com/viewer?a=v&q=cache:v1wzR2Hr5vAJ:www.cybersaa.org/2007_Outlook_PDFs/Press_AOC_07.pdf+%22cars+purchased+in+a+lifetime%22&hl=en&gl=us&pid=bl&srcid=ADGEEShKtRM4QN0oOcXdz4Chgn0SmL7T-vMvMmI32wOZOExMEiQ1oJGhvnlAvVTKcNBVGFtGMUTPtDDPsn3iVGzmVYwP7xTbKuEW5Xighqo8bZbTgvnRhzlKJz8BlsVMNEtIW8UFsdAz&sig=AHIEtbSDBlDmwEPRaDl39o2uf4RXG3HzQw>
2. Credit: Robert L. Hirsch, et al. *Peaking of World Oil Production: Impacts, Mitigation, & Risk Management.* U.S. Department of Energy, Pages 23-24, 2005; <http://www.hilltoplancers.org/stories/hirsch0502.pdf>
3. John German Testimony to Congress <http://energy.senate.gov/public/index.cfm?FuseAction=Hearings.Testimony&Hearing_ID=287981fa-c472-4862-857d-3913d416297e&Witness_ID=e2050b74-5cff-4f6f-9832-fec00f3ab26c>
4. One Billion Cars by Joseph B. White, *Wall Street Journal,* Apr. 17, 2006; <http://www.mindfully.org/Energy/2006/One-Billion-Cars17apr06.htm>
5. Safety Applications of Intelligent Transportation Systems In Europe and Japan, Department of Transportation, January 2006, Page 2 <http://international.fhwa.dot.gov/ipsafety/ipsafety.pdf>
6. *Transportation Energy Data Book,* 27th Edition, 2008, tables 2-11 and 2-12, Center for Transportation Analysis, Oak Ridge National Laboratory; 2008; <http://www.cleanenergycouncil.org/files/Edition27_Full_Doc.pdf>
7. Avego website; <http://www.avego.com>
8. Avego Powers VPSI Vanpooling http://www.avego.com/common/press/releases/Avego_VPSI_Press_Release.pdf
9. "Think outside the box on traffic" by Robert Behnke, *The Portland Tribune,* Jan, 6, 2006.
10. Smart Jitney/Community-Enhanced Transit Systems by Robert Behnke and Park Woodworth presented at the 2006 APTA Bus and Paratransit Conference; <http://faculty.washington.edu/jbs/itrans/itrans37.htm>
11. "U.S. Rethinks Electric-Car Fuel Ratings" by Sharon Terlep, *Wall Street Journal,* April 17, 2010.
12. Plug-In Vehicles: The First Great Fraud of the New Millennium by John Petersen, March 16, 2010 <http://seekingalpha.com/article/193847-plug-in-vehicles-the-first-great-fraud-of-the-new-millennium>
13. "Sequestering $C0_2$ in a Closed Underground Volume" by Professors Christine Ehlig-Economides and Michael J. Economides, Journal of Petroleum Science and Engineering 70 (2010) pages 123-130 http://twodoctors.org/manual/economides.pdf

Index